Computers in Manufacturing

David A. Turbide, CFPIM, CMfgE

Computers
in Manufacturing

Industrial
Press Inc.

Library of Congress Cataloging-in-Publication Data

Turbide, David A.
    Computers in manufacturing / David A. Turbide.
    203 p. 15.6 x 23.5 cm.
    Includes index.
    ISBN 0-8311-3033-4
    1. Computer integrated manufacturing systems. I. Title.
TS155.6.T86   1991
670'.285 — dc20   90-47446
                                                                         CIP

Computers in Manufacturing

Industrial Press Inc.
200 Madison Avenue
New York, New York 10016-4078

First Edition

First Printing

9 8 7 6 5 4 3 2

Composition by V & M Graphics, New York. Printed and bound by Quinn Woodbine, Woodbine, New Jersey.

Contents

Computer Integrated Manufacturing (CIM) is usually defined as the integration (interconnection) of the computers that are found in various places throughout a manufacturing company for the purpose of exchanging information.

As far as it goes, this definition is basically correct. The mechanics of CIM focus on the physical connection between systems and the sometimes difficult process of passing information over these connections from one application to another. The computers used in one area of a manufacturing company may be quite different from those used in other areas, and no one vendor is dominant throughout. Tying these incompatible systems together and enabling the exchange of information is definitely a challenge. To understand the dimensions of CIM, one must be familiar with the many aspects of the manufacturing enterprise, the business, terminology, and priorities of each functional area, and the technology of computers and communications.

Preface

The desire of various system vendors to make their products work together with those of the competition has only recently begun to overcome the myopic view that a single vendor can provide all of the solutions. Integration is much more feasible today than ever before, and is now within the reach of even small manufacturers.

While reading this book will not make you an expert in CIM, its intention is to introduce the major aspects of the technology and provide some insight as to how the various pieces can fit together. Beyond simple explanations of the technology involved in systems integration, the book also introduces the reader to the kinds of information that each functional area uses and maintains. Further, I have attempted to introduce some considerations about how the *functions* relate to each other, and thus go beyond the mere commonality of data elements. In so doing, I hope to raise the level of understanding and foster cooperation between areas that often don't work together very well.

This book is not intended to be highly technical. The description of technologies is purposely simplified to stress the important relationships and considerations while introducing basic concepts and terminology. When you have absorbed its contents, you should be better able to discuss CIM with people in all areas of the manufacturing enterprise and to better communicate with experts in the specific technologies involved.

Simply stated, manufacturing is the process of turning parts (or materials) into products. Although there is tremendous variety in the processes, materials, requirements, complexity, and terminology between manufacturing companies in different markets and product areas, there is enough commonality that manufacturing can be addressed as an industry in trade journals, professional societies, conventions and trade shows, lobbying groups, software packages, and books like this one.

1. Introduction

All types of manufacturing require input in the form of materials and/or components; all produce products that differ from the input materials because of the action of some kind of process that (hopefully) adds value; all require resources to perform this process, including people and usually machines or equipment of some sort; all are concerned with marketing, selling, and distributing the products in some manner; all involve some level of engineering and/or design; and virtually all face competition.

Manufacturers tend to classify themselves into two general categories: discrete and process. Discrete manufacturers assemble (put together from parts) or fabricate (transform materials or components through processes such as machining) individual products (discrete units—"each's"), whereas process manufacturers usually work with liquids and/or powders, and then process by mixing, blending, cooking, or by a chemical process such as refining. Process manufacturers often produce their products in large quantities, measured in units other than "each," and more often experience "yield" losses during the process. They usually develop products in a laboratory rather than a design department. Of course, these are very simplistic definitions and, in fact, many industries employ elements of both process and discrete production.

The label itself is not as important as the differences in emphasis of the management functions between the two extremes. Process plants (seldom called factories!) are often more highly automated than discrete production plants. The process environment lends itself to automatic production machinery that produces a large volume of the product at high speed. The machines tend to be special-purpose and are more difficult to change over from one product to another.

1

The raw materials used by process manufacturers are often commodity items like plastics, organic compounds, water or solvents, dyes, and chemicals. The material costs are typically a small fraction of the cost of goods, so raw material inventory investment is not as big a concern as it might be in other industries. In addition, shortages are deadly since the expensive high-speed machinery cannot run without the proper materials, so material availability is much more important than minimizing inventory investment. Since keeping the machinery in productive use is the major concern, the automation (and management) dollar goes into machine controls, quality monitoring, and day-to-day production management tasks at the expense of sophisticated long-range material planning tools, or design and drafting systems (generally not needed).

Another distinction between manufacturing environments is continuous versus batch production. Both situations exist in discrete as well as process industries. In continuous production, the work in progress moves from one operation (production activity or process step) to the next as each item or small group completes the operation. In batch production, an order (production quantity) will complete an operation in total before moving on to the next step. Continuous production is usually reserved for high-volume products, and tends to utilize material handling equipment such as conveyors, belts, robots, or trays to move the work from one station to the next.

Continuous production results in relatively short production lead times, and the need to coordinate production rates so that the work can flow smoothly from one station to the next. Most process plants are continuous, as are many high-volume discrete plants. Many continuous plant processes use specialized automated equipment, but this is not a requirement.

Despite these kinds of differences, all manufacturing companies face the need to develop some kind of plan or schedule, bring in the materials necessary to support the plan, have the resources available to produce, and manage the sale and distribution of the product that results. Most manufacturers have a need for design (or research), engineering, change or configuration control, and industrial (process or production) engineering functions. All manufacturers must manage the production process itself including resource utilization, activity reporting, and quality monitoring.

Computer Integrated Manufacturing (CIM) is the recognition that there are computers in use in three distinct areas of the manufacturing business; that the systems in each area contain and use information that can be of benefit if made available to the other areas of the business; and has as its goal the integration (transfer and interrelation) of this information where a benefit could result. (See Fig. 1-1.)

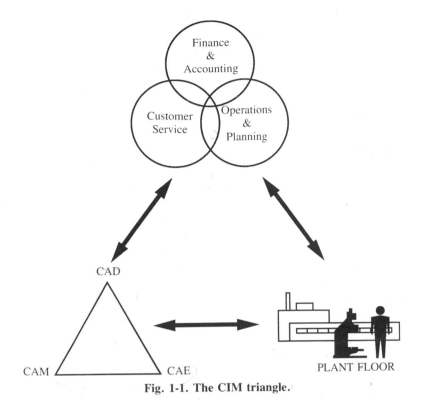

Fig. 1-1. The CIM triangle.

The three areas of automation are: business and planning, design and engineering, and the plant floor. Historically, the application of computer technology to the needs of each of these areas has proceeded with virtually no recognition of the existence and needs of the other two. Therefore, we have had different vendors developing distinct, and for the most part incompatible, solutions in each area. Only recently has there been significant advancement in the ability to interconnect these various systems. In parallel with the development of hardware (and communications system) interconnection capabilities, software vendors are becoming involved in data exchange and the integration task to make use of the new connections.

Business and Planning

In the business and planning area, the systems address the general business functions of finances and accounting, customer order servicing, material and production planning, and operations (production and purchasing). Integration within this area is well advanced, and the general term for these applications is Manufacturing Resource Planning (MRP II). These "front office" functions typically reside on a mainframe or midrange computer, although small systems are beginning to make inroads with the growth in capabilities of "personal" computers. From a system standpoint, the business and planning tasks are primarily data management functions. The computers used here must be adept at multiple concurrent user access, and must be very good at storage and retrieval of large volumes of information.

MRP II handles the planning and management of materials, production, and purchasing. Production requirements and priorities are logically linked to plant-floor activities and should be electronically tied together as well. All activities are coordinated with overall company objectives (making and shipping products on time) and accounting functions are integral to the operational functions. MRP II functions rely on product and process definitions that are stored in its database. Much of this data is developed and maintained in the engineering systems, and there are some obvious efficiencies to be had by linking these functions together.

Design and Engineering

On the design side, Computer Aided Drafting and Design (CAD or CADD) systems have evolved down in price dramatically over the last five years. In the early 1980's and before, a high-function CAD system required a mainframe computer and expensive displays and output devices (plotters) with an entry price in the hundreds of thousands of dollars. The breakthrough in CAD came as a result of the development of "workstations," powerful small computers capable of supporting the demanding tasks of handling mathematically described graphic images. While workstation computers are the dominant platform for CAD, there are now some rather widely used PC-based products with some impressive capabilities on today's larger, faster PC's. The system task for CAD is calculation speed, known as "number crunching" in the business. Workstations are compared based on their ability to process so many millions of instructions per second.

Mainframe CAD is still widely used, especially in larger companies with many CAD users. The large systems provide superior processing capabilities as well as large storage facilities and multiple user support for convenient access to a centralized "library" of drawings, standard images, and analytical functions. Once an object is mathematically defined in the CAD system, it can be analyzed using

Computer Assisted Engineering (CAE) functions that have access to the drawing data file. These functions can display "solid" looking representations of the object, simulate the motion of moving parts, allow clearance checking of moving parts, analyze the effect of stress and loads, and calculate the heat conduction and dissipation characteristics. All of these functions require powerful computer calculation capabilities.

Also, based on the description of the object, a supplemental program can generate machine-tool instructions which can be passed to a production machine used to manufacture the object. The process of generating the instructions and using them to produce the object is called Computer Assisted Manufacturing (CAM).

Plant Floor

Automated controls for production machines originated with manually created instruction "programs" used by weaving looms during the days of the Industrial Revolution. Early computerized machines were given their instructions on paper tape and were called Numerical Controlled (NC) machines. The fixtures and tools were set up manually, often requiring hours of preparation before running the programmed sequence. Computerized Numeric Control (CNC) is associated with more flexible machines which are capable of multi-axis movement, more intricate instruction sets, and sometimes automated tool changes.

The capabilities of CNC machines are limited in part by the amount of memory available for instructions, and by the time and effort required to load the program. These problems are overcome with Direct (or Distributed) Numeric Control — DNC. With these machines, the program instructions are stored in a separate computer which is not a part of the machine itself but is attached to the machine through a communications channel. This computer can be a PC or a workstation, and may store and "download" the programs for a number of machines.

Also on the plant floor, there is often a need to collect production information for quality monitoring purposes and to support the business and planning systems (job tracking, efficiency, cost accounting, payroll). Sometimes data capture equipment is attached directly to the machine or could be built into the controller. Production information is most commonly collected through shop-floor terminals, bar-code data collection equipment, or hand-held terminals, if not manually recorded.

Advances in plant automation include flexible machining centers, work cells, programmable robots, and automated material handling equipment. Each of these can be set up to reliably perform repetitive tasks, and will be discussed in more detail in a later chapter.

Machine controls operate in an environment that is quite different from that of computers that interact with humans. In a human-based application, the system presents information on a screen and waits for the human to respond. When the operator completes an entry, the system's action then takes place in a burst of activity whereupon the system's next response is displayed and another wait period ensues.

With controllers, there is a more continuous exchange with the other device. The controller issues an instruction, waits for a response (sometimes), then issues the next instruction. In many cases, the controller is monitoring a number of inputs (machine outputs) simultaneously and is comparing the input to a set of expected conditions. Based on the information received, the program can initiate various responses such as a command to move a cutter or reposition a workpiece.

These activities take place on a time scale in which humans can't directly participate. The communication between the machine and the controller typically happens many times each second. Programs for these applications deal with very small pieces of information, sometimes only a voltage level or a single pulse, and the instructions for the machine are often a single character or "word." Control computers are special-purpose microprocessors or microprocessor-based programmable units.

Integration

When a new product is conceived, that product might be drawn and specified on a CAD system. Using the CAD data, the properties of the new item can be tested for strength, movement, etc., with supplemental programs, probably on the same computer. Once the design is finalized, another program can create the machine-tool program with which to fabricate the part. Item information and material specifications can be passed to the business and planning system to coordinate the acquisition of materials needed for production.

After industrial engineering has defined the process and planning has authorized production, the business system will assist in setting the production schedule and will notify the plant floor as to how many to make and when. Materials are reserved, then issued with shop instructions, and manufacturing begins. The machine program is downloaded from the engineering system at the appropriate time, and production information is gathered for Quality Assurance and to update the business system.

When completed, the part is moved to stock where it is available (just in time) for use in the assembly or product that needs it.

In most companies today, no such efficiencies are available. It is more typical to have the design prepared by hand or on a stand-alone CAD system. The testing of the part is done with a prototype, and revisions are manually added to the drawing. Part descriptions and material information are manually entered into the business system, and machine instructions are developed on the plant floor.

When it is time to produce the part, the program is retrieved from a file on tape, or perhaps downloaded from a PC used exclusively as a program storage facility. Production information is recorded on time cards or reporting tickets and keyed into the business system. Separate descriptions of the part exist in the CAD system, the machine instructions, and in the business system. Seldom are they coordinated when any one of them is changed.

Past Directions

Historically, developments in the application of computer technology have advanced in the different areas of the manufacturing enterprise at different rates and using different technologies. In truth, the requirements of the three areas are different enough that there could probably not have been a general-purpose processor powerful enough and flexible enough to satisfy all areas even if there had been a strong motivation to do so.

In machine controls, the early task was merely to be able to accept a rather short list of instructions and execute them in sequence. No great intelligence was required, no data handling, no demanding input–output tasks; just ruggedness, reliability, and a very modest amount of storage to hold the instructions. As these systems have become more sophisticated, these basic requirements have expanded to process more instructions and have more ability to accept input from the machine and respond to these stimuli. More complex programs can be generated or downloaded, and many machine controls also include graphical output and programming-assistance screens as well as data-logging and communications capabilities. In comparison to today's business systems, these programming requirements are so fundamental as to be almost trivial, but this is not the strength of process control computers.

Control computers must accept and process many inputs and outputs very quickly. The response (cycle) time in these applications is measured in milliseconds or sometimes microseconds. When dealing with human operators, the difference between half-second response and one-second response is barely noticeable. In a machine controller, many functions can be performed in that one-half second.

Microprocessors, sometimes called "computer on a chip," constitute the dominant process controller computer type. These smart integrated circuits are relatively inexpensive, fast, and increasingly more capable. Microprocessors are teamed with memory modules (integrated circuits) to provide program storage and work space. Various devices to handle the input and output tasks such as signal conditioners, analog-to-digital converters, line drivers, and display modules or interface devices are used to manage the communications with the outside world. Often, all or most of these components can be mounted on a single circuit card for convenient packaging and common usage in a number of machines.

Newer controllers have improved capabilities to communicate with other computers. These controllers can accept instructions from another computer either during a set-up operation (computerized numerical control or CNC) or while controlling the process (Direct or Distributed Numerical Control, or DNC).

Business systems must be designed to handle lots of information efficiently. Storage requirements are very large, and programs can be long but typically do not require extensive calculations. The primary tasks are to store and retrieve large amounts of data and be able to manipulate them: sort, summarize, and display. Business systems typically have a large number of interactive users (those having a two-way exchange with the machine) so these systems must be built to communicate effectively with multiple simultaneous users. Finally, business systems should include data management functions or software (called Database Management Systems or DBMS) to allow flexible access to the stored information.

Business systems used to require mainframe computer capabilities. As smaller systems (originally called minicomputers but more recently referred to as "midrange" systems, after the traditional lines between mainframes, minis, and micros began to blur) became larger and more capable, more and more business systems began appearing on midrange hardware. As the price of all size systems declines at the same time that power and capacity increase, business systems are beginning to become available on microcomputers and microcomputer networks.

Engineering systems are optimized for processing mathematical calculations quickly. The graphic systems used in the design and drawing functions really describe the lines and shapes of the images as mathematical equations. Manipulating these shapes, displaying different views (from different sides or angles), changing the size of an image or rotating it all require a tremendous number of calculations. For the system to be responsive, these calculations must be performed quickly (many calculations per second).

CADD systems often are single-user oriented. Multiple-user CADD is usually separate "seats" sharing a central file storage function but operating quite independently. Other engineering tasks, such as analyzing the figure or simulating motion, are equally or even more demanding from a calculation sense, so engineering computers, more than anything else, must be fast.

The dominant computer type for engineering is the "workstation." Workstations fall somewhere between microcomputers and mainframes on the power, cost, and size charts, but differ from midrange systems in that they are designed for single-user demanding applications rather than less mathematically demanding tasks but much higher volume interaction and data management tasks that the midrange supports. While there are a number of manufacturers of workstations that encourage the use of proprietary (unique and copyrighted) operating systems,[1] there is a movement toward standardization on AT&T's UNIX operating system and its derivatives.

There are some very successful CAD systems which run on personal computers (PC's, the most commonly used name for microcomputers). The new generation of PC's typified by the larger IBM Personal System 2 models and other systems using Intel's new, more powerful processors (80386, 486) indicate that PC-based CAD will be a major force in the market before too long.

In conclusion, it is becoming increasingly more difficult to characterize computers and draw distinctions between "types." The hardware technology is changing very rapidly, and "small" systems are getting larger and more powerful all the time. Applications that, a few years ago, could only run on expensive mainframes now are available in the midrange and even on PC's. There is also a movement toward standardization not only in operating systems but also in programming languages, communications protocols, and interface specifications (connections between systems).

---

[1]An operating system is a set of instructions that act as an intermediary between the user programs and instructions that the processor itself understands. Most companies have developed their own unique operating systems which are incompatible with all others. User programs written for one operating system typically cannot be moved or "ported" easily to another operating system. There has been some movement toward standardization in workstations toward AT&T's UNIX (and similar systems such as IBM's AIX, XENIX, and others) which is now available on many personal computers and even midrange and mainframe systems. In personal computers, Microsoft's MS-DOS (similar to and compatible with IBM's PC-DOS) is a de facto standard.

All manner of computer equipment vendors are beginning to recognize that they are not alone in the factory. They have come to accept that they will not be able to displace all other vendors and provide for all of the users' needs, and that to survive with their piece of the business intact they must learn to work with the other vendors' systems. Every trade publication is filled with announcements that this company or that company has established a CIM division or has purchased an integration company or announced products that communicate with others' equipment.

The time for CIM has arrived and it is none too soon. In today's increasingly competitive world, companies must take advantage of every facility available to increase efficiency and flexibility and to reduce costs and better control production. By taking better advantage of the computer power already in use in the plant, through integration, we hope to enhance our decision-making capabilities by being better informed of conditions and of the impacts of our decisions.

Integrating

The foregoing discussion notwithstanding, CIM is not just physically connecting together the various computers within the enterprise. CIM is an overall approach to management of the entire company and the data and information that is a part of the everyday operation of that company.

To implement CIM is to recognize that all areas of the company are interdependent. From initial concept through design, engineering, planning, production, customer service, and after-the-fact analysis, nothing happens without having an effect on what comes after, or without being affected by what came before. To take a CIM approach to management is to recognize these relationships and to use them to advantage.

The most recognized aspect of this approach is known as Early Manufacturing Involvement or Design for Manufacturability. As the name implies, it is getting the design department to listen to the people who will eventually be asked to produce the product. In this way, the design can be influenced at an early stage, where it is easiest to change, to make the product easier (more efficient) to build. Too often, engineering and manufacturing don't talk to each other except to point fingers and throw invectives during "How did we get this screwed up?" meetings.

Many times, a design can be modified slightly, without changing the function or outside appearance, to allow significant savings in manufacturing. This is especially important when automated processes are involved on the plant floor. While robots can do marvelous things, they are more limited in "manual" dexterity and flexibility than humans. A minor redesign of a component to include a chamfered edge, for instance, that is more forgiving of minor misalignment in assembly could increase a robot's success dramatically in day-to-day operation. While a design engineer might not consider this significant to the overall functionality of the product, a manufacturing engineer or technician might well recognize the benefits of the change immediately on studying the drawing.

Putting the impact of a design change in perspective, think of the cost of a simple drawing change compared to a change after production has begun which would include retooling, reprogramming machines, retesting (or certification), scrapping or remachining any on-hand inventory of parts, production disruption during the change process, documentation changes, etc., and you can begin to see the advantages of early involvement. Design for manufacturability has nothing to do with computer integration. It is the direct interaction of people from different business areas that makes it work.

There are some computer tools, of course, that can assist in this area. First on the list is Computer Aided Design that makes drawing modification so easy and efficient. Next are the Computer Aided Engineering tools that can be used to simulate the effect of changes, test the durability and other physical characteristics of the new design before a single chip flies, and can also simulate some of the production processes such as checking for possible interference as parts are assembled.

One of the computer-based tools that is becoming more important in design for manufacture is Group Technology (GT). This is nothing more than a classification scheme that recognizes the similarities between parts. By grouping items together according to physical characteristics such as size and shape, and according to the production processes they employ, design can more readily recognize existing designs that can be reused or adapted to new applications, thus saving design and engineering time as well as reducing inventory and saving money.

Group Technology is generally considered to be a tool for engineering but is most effective when used to make production more efficient. Should it then be considered production's responsibility? If engineering perceives Group Technology as something that they must do for production's sole benefit, where is their motivation to embrace it and make it work? And who is responsible for defining the coding scheme, assigning the codes to all items, and maintaining the integrity of the system?

These are not trivial questions. The success of any of the integrated systems and the techniques discussed in this book is totally dependent on these kinds of considerations. Computers will do what they are programmed to do. People will do what they are motivated to do. Any implementation project must consider how the changes will be viewed by the people affected and whether or not the current environment will support the desired objectives. This especially includes the way people are motivated. Incentives must directly encourage the desired behavior.

The points made about early involvement and group technology apply to all other interactions across functional lines. CIM requires that information be shared for the common good. It requires that each department and each employee feels that there is no reason to hold anything back and, indeed, feels a strong need to communicate freely and openly with the other areas of the company. The biggest impediment to implementing CIM can easily be interdepartmental rivalry or jealousy.

Often, this rivalry develops from competition for resources. Most companies exhibit a "personality" that favors the engineering, marketing, or production side. The bent of the company will reflect the key factor(s) in that company's past success. If the company is in a technical business, engineering is likely to be the driving force of the business. It is therefore also likely that high-ranking executives came from engineering and they are more likely to approve capital expenditures requested by the engineers than those requested by other departments. The same is true for companies dominated by marketing, production, finance, etc., as they tend to allocate resources more liberally in support of the area(s) that have been key to past success.

Whether or not there is justification for these feelings, the rivalry and competition for resources tends to discourage the free flow of information between functional areas. No department is motivated to make another look good at what may become, ultimately, their own expense. As new information bridges are erected in a CIM effort, it is important to recognize the proprietary feeling that people are likely to have toward their "own" information, and it is important to sell the benefits to each user, not only for themselves but also for the company as a whole. Fortunately, effectively implemented information management systems tend to reward the providers of information with a return that exceeds the value of the input. Through the synergistic effect of combining their data with data from other sources into an integrated system, each function that contributes should receive back far more value than they could have received in a stand-alone application. This concept must be "sold" to the users.

Another impediment to implementation of any new management system is an employee's motivation system. People tend to do what is rewarding (pleasurable) and avoid what causes pain. The motivational system that is currently in operation in the plant is a reflection of the priorities and directions that currently exist in the company as well as a derivative of those from the past. When new systems and procedures are implemented, it is important to evaluate the motivational environment to ensure that the desired behavior is properly encouraged.

If the new information system is designed to provide plant-floor personnel with dynamically generated priorities tied to current demands and schedules, then it is important that production heed these priorities for the system to be effective. If these same production people are motivated based on the number of units produced per month or daily production counts, they are not tied in to the system's priorities. When it is time to select the next job from the available queue, he/she will select the one that provides the best result for his/her incentive system, not necessarily the one that the system says is most critical. He/she may well select the lowest priority job next because it gives him/her the best numbers for the day or month.

One of the least recognized and often most difficult areas of system implementation is identifying where employee motivation is not in line with the objectives of the new system, and then correcting the motivational system. Often, this can spell the difference between mediocre results and truly outstanding improvements.

Industry Trends

Manufacturing today, in nearly all markets, is far more competitive than ever before. Competition tends to be more global and promises to become even more so. With emerging countries rapidly developing manufacturing capabilities and the reduction of trade barriers worldwide, most manufacturers are now in global markets with global competition.

At the same time, product cycles are getting shorter, at least in part due to the increased competition. Rapid advances in technology also render products obsolete far more quickly than before.

For these reasons and others, we no longer have the luxury of ample time to develop new products and bring them to market on a leisurely schedule. Anything that we can do to reduce design and engineering time, and get products to market more quickly, enhances flexibility and competitiveness.

Twenty years ago, the conventional wisdom was that higher volume results in lower unit cost. While the concept of economies of scale is itself still valid, the realities of the market place definite limits on this logic. When considering fixed and variable costs, it is simple to see that when a larger production quantity absorbs a fixed cost, the impact on the individual units is reduced. If, however, a large number of the goods produced end up as obsolete inventory because of the introduction of an improved product or a lower-priced competitor, then the savings are illusory. The result is a trend toward smaller "economic" production quantities. This can be achieved only by reduction of fixed costs. Fixed costs include development, engineering, and production set-up costs. Anything that helps reduce these costs improves competitiveness by allowing us to meet market needs more precisely. Fortunately, the things that help reduce costs often reduce the time required as well, giving us more flexibility.

With smaller production lots, shorter product cycles, and intense competition, companies are striving to differentiate their own products from those of the competition. While the "economies of scale" theories of the past tended to reduce the selection of products available because the price of entry is high for high-volume production, the foregoing trends increase diversity. In an effort to avoid direct head-to-head competition, and taking advantage of the flexibility offered by the new technology, companies are now offering an ever-wider variety of products aimed at smaller "niche" markets. This often allows higher margins and discourages direct competition because of the limited sales potential of the precisely defined target customers.

All of these trends are related by the interaction of increased competition and changing market conditions combined with the advances in technology and management that allow the outlined changes in the production realities.

Manufacturing management has been constantly searching for new tools and techniques that will improve competitiveness, increase flexibility, reduce costs, and help manage resources more effectively. This search has led us through a bewildering array of new software products, theories, and systems all designed to "save" existing manufacturers from the competition. Sometimes, it's difficult to distinguish between what is really new and different and what is just hype or merely a refinement of an existing technique.

This is especially true in the business and planning area, but applies in engineering and plant operations as well. For example, MRP II, Manufacturing Resource Planning, is a widely used system-based approach to planning and managing the resources of a manufacturing company. Several years ago, "Just-In-Time" was widely touted as the newest and best method for becoming and remaining competitive. People started to ask (and some are still asking) "Should we install MRP II or JIT?" or "Should we scrap our MRP system and use JIT instead?" In fact, JIT is not a system at all but a management philosophy that can be applied in any company, whether MRP is in use or not. Actually, MRP II is often used as one of the tools to achieve JIT objectives.

JIT is a focus on identifying wasteful practices and taking action to reduce or eliminate waste. JIT is a continuous improvement program that often includes many of the tools and techniques discussed in this book including MRP and MRP II, quality control techniques such as statistical process control, and many types of factory automation.

The real danger in the development and promotion of new techniques and theories is that the attention is drawn to the technique and distracted away from what is really important — producing a quality product on time and at an acceptable cost. In addressing CIM, we must be aware of that danger and keep clearly in mind that no clever technique, fascinating theory, sparkling new computer, lightning-fast communications system, or expensive new machine is worth anything unless it directly supports the objectives of the company and helps it to become and remain competitive and profitable.

With that in mind, let us explore the application of computer technology in the three main areas of the manufacturing enterprise: design and engineering, business and planning, and the plant floor. We must discuss how these systems can be made to "talk" to each other, but let us not forget that the people must also talk — that any and all procedures, as developed, must include consideration for the impact that the activity will have on all other areas of the company, not just the task at hand.

A manufacturing company is an entity in which all of the parts are interdependent, and which is in existence for a single reason: to make a profit by producing and selling a product.

Chapter 1 Review Questions

1. Characterize discrete versus process manufacturing. Is this distinction important? Why or why not?

2. What are the three areas of automation in a manufacturing company? Why should we be concerned with their integration?

3. What is the common or related information between each pair of "areas of automation?"

4. What are the primary characteristics of computer systems developed for each of the areas of automation?

5. Outline the evolution of computer types in each of the three areas of automation (mainframes, microprocessors, PC's, etc.).

6. What are two (people-related) impediments to implementation of CIM systems?

7. What business/market trends are driving the move toward CIM?

8. Which is better, MRP II or JIT?

Applying the capabilities of the computer to assist in the preparation of a graphical representation of an object offers many advantages, not the least of which is the precision with which an image can be created. Dimensions can be stated in the units desired and the drawing will be a very accurate representation. Circles are always round, intersections are clean and precise. In addition, it is easy to imagine the productivity gains that are available if one can simply recall an existing drawing and modify it, compared to the laborious task of recreating the entire drawing from scratch.

2. Computer Aided Design and Engineering

While these are obvious benefits and they are, in fact, reasons that computers were first assigned these tasks, there is far more utility in the fact that modern CAD systems represent the physical characteristics of the object depicted, in mathematical form. The visual presentation is only the beginning. The mathematical descriptions can be used to develop what the properties of the object would be if it were to be built, and to test (simulate) its behavior under different conditions. The imaging process offers time and labor savings; the resulting mathematical description opens the door to an ever-increasing range of engineering capabilities.

CAD originally meant Computer Aided Drafting (or Drawing). In recent years, the definition has evolved to Computer Aided Design which implies more far-reaching capabilities. Sometimes the acronym CADD is used to include both design and drawing.

Computer Assisted Engineering (CAE) is the use of the mathematically described design information to explore the properties of the object with computer programs. Computer Assisted (or Aided) Manufacturing (CAM) includes the development, storage, and transfer of machine instructions such as tool-path data to the factory floor. Most definitions of CAM include the use of computer control in the plant (NC, CNC, and DNC) as a part of CAM. These topics are covered in a later section of this book beginning with Chapter 7.

Early CAD

The first CAD systems were intended to supplement or replace the drafting board in producing a precise, dimensioned drawing of an object. The advantages, as mentioned in the first paragraph, included the ability to specify a line length, the radius of a circle, or a set of coordinates, and to be confident that the resulting picture would be perfectly proportioned. The problem with these early systems was that they were not very easy to use. The primary input method was through the keyboard, and it took considerable skill and a lot of patience to render a useful image.

Once the necessary skills and a library of drawings were established, however, great savings were available where existing drawings could be modified for new uses. Reuse of parts of an existing drawing can yield up to ninety percent savings in the time required to complete the drawing for a new part.

Several developments in computer hardware and software revolutionized the usability of CAD systems. On the hardware side, the first was the development of the light pen which allowed the operator to point to a place on the screen rather than describe the spot using keyboard-entered coordinates. More recently, the light pen has given way to a hand-held device that is moved over a flat surface to indicate the desired direction and amount of movement to apply to a pointer displayed on the video screen. This device is variously called a "mouse" or a "puck," it is less expensive than a light pen implementation, and is becoming familiar to millions of personal computer users as a common accessory available on home systems. Other pointing devices in use include the trackball and the joy-stick, familiar to video-game players everywhere.

Software has also evolved to include many conveniences, particularly windowing and pull-down menus which assist the operator in translating his/her thoughts to the image. Lists of options (menus) displayed on the edge of the screen can be accessed by pointing to the desired function (moving the on-screen pointer via the mouse); and if there are other choices to be made to complete the command, a list of these is brought in (pulled-down) onto the screen for selection. Windowing refers to the ability to designate a subsection of the display screen for another purpose. As an example, a portion of the drawing in the display can be marked using the pointer and menu commands, and moved to a designated display area where it can be manipulated (magnified, rotated, etc.) while the drawing in progress remains undisturbed, although a portion of it may be obscured temporarily by the "window." The window is also useful because it allows other drawings, look-up tables, command lists, etc., to be displayed in a portion of the screen while the drawing is still at least partially visible.

Other developments that have made these systems more "usable" include high-quality color image display screens, more powerful software features, faster processors for better response time (less waiting), and extensive collections of standard shapes and common objects that can be brought into the drawing as needed.

Making a CAD Drawing

Making a drawing on a computer screen is quite different from putting pencil to paper. With pencil in hand, the command process is automatic, that is, you don't have to tell your hand to move up two and one-half inches, just think of the direction and the length, and the hand can be made to move as desired. In a CAD system, you have to tell the system in words, commands, and by pointing what it is you want done. The system does all of the drawing in response only to your explicit instructions. The CAD programs are capable of producing lines and figures of the designated size, shape, and length and in the position that you designate. Using a few basic forms such as a straight line or a circle, plus some functions such as scaling (designating the size) and rotation (angle), the image is built element by element. Perhaps the most useful feature of the CAD software is the ability to add *and* subtract the shapes. For example, to draw the shape in Fig. 2-1, start with a rectangle of the desired size (A), place a circle in the proper position (B), and subtract B from A.

When we think of drawing an object, it is typical to think first of a two-dimensional (2D) representation such as the one in Fig. 2-1.

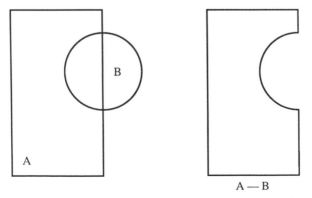

A — B

Fig. 2-1. Using two elements (subtraction) to produce a shape.

We have only defined here the appearance of the object from one side. Obviously, this is not enough to describe how to make the object because we don't know how thick it is or whether there are any features on the other surfaces. It takes three such drawings, generally, to fully describe the three dimensions of the object. The normal convention for engineering drawings is front, top, and right-side views. Traditionally, these three 2D drawings are placed in three quadrants of a page to make up an orthographic representation. The fourth quadrant can be used for an oblique, perspective, or isometric representation of the object (Fig. 2-2).

In 2D drawing, only the boundaries (edges) of the object are defined. The ability of a CAD system to link the two-dimensional information together from several views, such as on an orthographic drawing, and produce representations of depth is referred to as two and one-half dimensions (2½D). 2½D systems are less expensive than 3D, require relatively little extra computer power over 2D systems, and can provide useful capabilities for describing simple shapes such as sheet metal parts, and for representations of assemblies where precise manufacturing information is not required (no intention to use the drawing to perform CAE or CAM functions).

Three-dimensional (3D) CAD systems provide the maximum capability to fully describe an object in all its physical form including all outside surface features and internal structure. There are three varieties of 3D CAD systems: wire-frame, surface modeling, and solids modeling. Wire-frame is most similar to 2D and 2½D systems in that points and edges are represented. The figures displayed are basically transparent and are therefore sometimes difficult to interpret. To emphasize the front surfaces and deemphasize the ones behind, different colors can be used, certain surfaces can be overlaid with a grid pattern, and "hidden" edges and surface grids can be dimmed or removed altogether.

Surface modeling systems have the ability to display a seemingly solid surface, thus obscuring edges and lines that would ordinarily not be visible in the physical object from that view. Surface images are particularly useful in showing contours especially with the use of shading techniques.

There are two types of solid systems. The first, boundary representation, is an extension of the 2D and 2½D processes of defining the shapes by specifying the edges (boundaries). Constructive solids modeling, however, is literally another dimension in object definition and representation. Unlike the use of basic shapes and line specification to build a definition of the object's edges as in other techniques, solids representations are built from three-dimensional objects such as cubes, cylinders, cones, etc. These basic objects are called primitives, and are positioned, scaled, added, and subtracted to derive the desired shapes.

Constructive solids modeling requires considerably more computer power than the other techniques but provides the most complete representation of the object and contains the full mathematical description needed for computer-generated engineering analysis and production machine program generation. Given the proper equipment and training, CAD with a constructive 3D system can produce better designs, faster than other methods at least in part because the design process more closely simulates the actual fabrication; i.e., subtracting a cylindrical shape from a cube is a lot like drilling a hole in a block of metal. Two-dimensional drawings and various views (oblique, perspective, isometric) can be generated from the 3D model easily.

In addition to the primitives (basic geometric figures), most CAD systems are available with an extensive library of commonly used parts and shapes. These figures can be brought into the design-in-

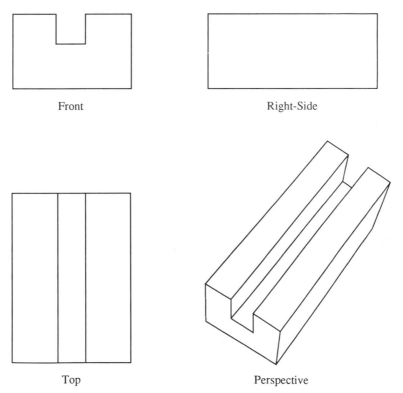

Front   Right-Side

Top   Perspective

Fig. 2-2. Conventional engineering drawing including a perspective representation.

progress very easily (designate position, scale, orientation, and bring it in), adding an entire element of the design at once. Common libraries will include nuts, bolts, brackets, gears, etc. Systems also provide a capability to allow the user to add figures to the library.

When CAD designs are constructed, they are typically built in layers. The layering is a way to subdivide the design into a series of partial definitions that can be used together, alone, or in various groups for different purposes. Using a 2D or wire-frame example, the first layer might contain the outline, center lines could be the second layer, dimensions on the third, cross-hatching on the next, notes next, then part numbers, titles, etc. This organization allows the user to display a screen image or produce a drawing with only the desired elements. The layers can also be easily displayed in different colors for increased clarity. Each layer is stored in a separate portion of the file that includes all layers for the drawing and carries a unique identification that allows retrieval for display or analysis.

Uses of the CAD Design

Once the geometry has been defined in the CAD system and stored in a computer file, it can be made available for other uses. Obviously, a similar part can be designed easily by starting with an existing design and adding only the changes (differences). Identifying the candidate for modification, however, is not always a simple task. In a large company or one with thousands of drawings, retrieving the closest design or even identifying a group of drawings that are similar to the new part is often a daunting task.

Many companies have developed a part numbering system that identifies many characteristics of the part. These systems, referred to as "significant" numbering schemes, tend to result in very long, complex part numbers. As an example, a part number such as "3127-1006-HT-016/24-257J,012" designates the part family (3127), material type (Carbon Steel AISI # 1006), heat treated (HT), thickness (016 Gage), coil width (24 inch), die number (257J), and revision. While the number itself now identifies (to a knowledgeable person) exactly what the part is, keying such a complicated number into a computer, repeatedly and accurately, can be quite difficult. Overall, part numbers are the most frequently entered data items, and the inclusion of special characters such as a hyphen, commas, the slash mark, and significant positions requiring spaces and/or zero-fill can dramatically slow down entry and increase the frequency of errors.

Another major problem with significant numbering systems is

that, no matter how comprehensive the layout, something usually comes along that does not fit into the scheme. When this happens, the system breaks down and chaos results. In a manual storage and retrieval environment, the part numbering clues can be of great assistance, but computerized databases don't care what the numbering system is. All parts are treated alike and other fields are provided for grouping and sorting.

A formalized grouping philosophy based on the physical characteristics of an item is Group Technology (GT). The GT coding scheme is used to classify the part or product and assign a GT code to it according to a predetermined set of codes that relate to defined physical or process characteristics and can be used to organize part descriptions (drawings and computer-stored geometric definitions) to assist in the retrieval of parts, and can also be used to group parts according to the manufacturing processes that they employ to assist in process planning. A discussion of the application of GT on the production side is included in Chapter 3.

There are several systems in use for GT, at least two of which can be purchased as computer software. The development of a GT system for a particular company, however, is usually an adaptation of one of the existing techniques.

One of the earliest and best-known systems was introduced by H. Opitz (University of Aachen, West Germany) in his book *A Classification System to Describe Workpieces* (Pergamon Press Ltd., Oxford, England). The Opitz system consists of a five-digit number followed by a four-digit number, with an optional four-character suffix.

The first five digits use a hierarchical structure to define the form of the item (form code). The first digit designates part class: zero through five are for rotational parts, and six through nine are for nonrotational parts. The second digit is for external shape; the third is for internal shape; the fourth is for surface characteristics; and the fifth is for auxiliary holes and gear teeth. The form code portion of the Opitz system defines general design characteristics that can be applied to the design retrieval grouping task. See Fig. 2-3.

The first five characters of the Opitz system (or the equivalent coding in another GT system), therefore, can be the basis for retrieval of existing drawings which can potentially be reused to save design time. If an existing part can satisfy the requirement, not only could design time be reduced to nearly zero, but existing inventory could be used to satisfy the requirement, lead time is saved, and fewer inventory parts are needed.

Reducing the number of unique parts is a key contributor to success in a Just-In-Time approach to manufacturing. Fewer parts

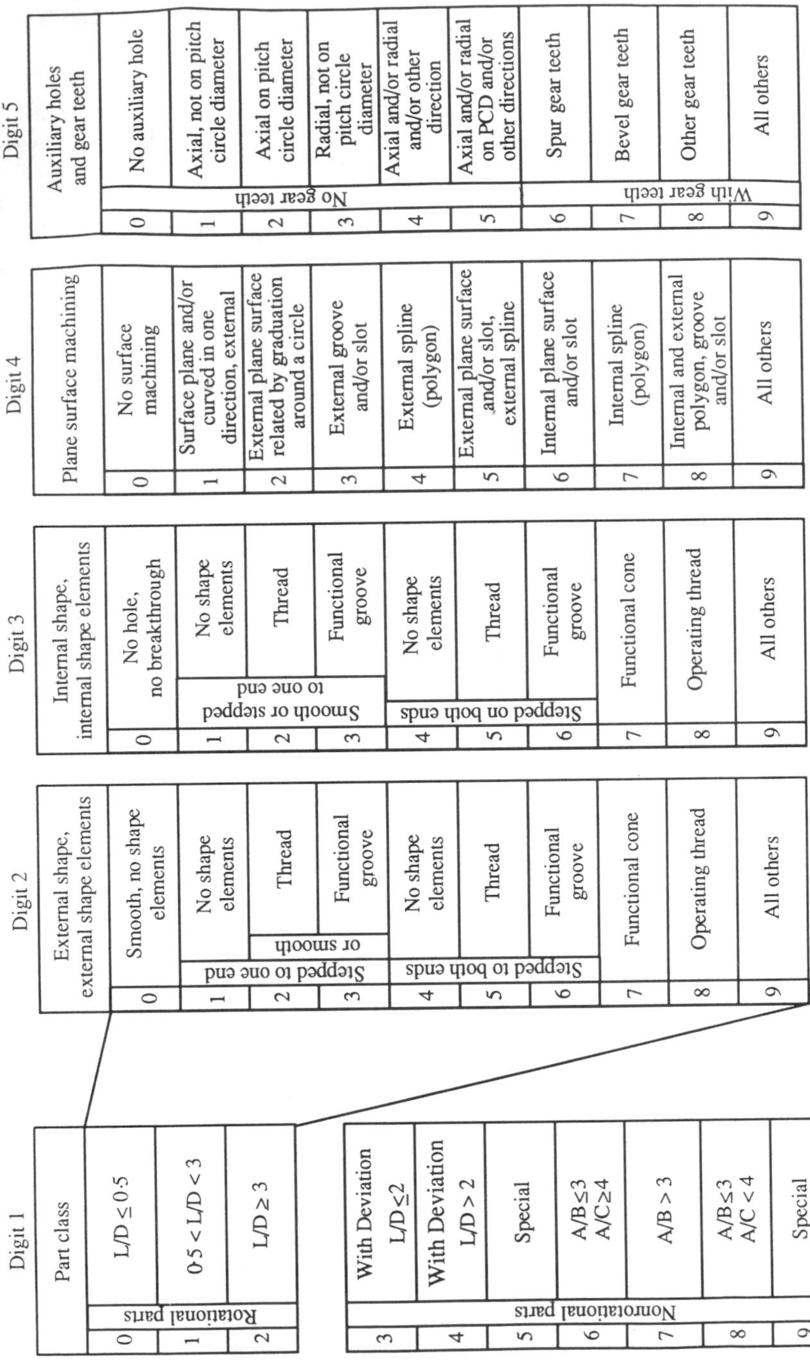

Digit 1 — Part class

Rotational parts	0	L/D ≤ 0.5
	1	0.5 < L/D < 3
	2	L/D ≥ 3
Nonrotational parts	3	With Deviation L/D ≤ 2
	4	With Deviation L/D > 2
	5	Special
	6	A/B ≤ 3, A/C ≥ 4
	7	A/B > 3
	8	A/B ≤ 3, A/C < 4
	9	Special

Digit 2 — External shape, external shape elements

	0	Smooth, no shape elements
Stepped to one end or smooth	1	No shape elements
	2	Thread
	3	Functional groove
Stepped to both ends	4	No shape elements
	5	Thread
	6	Functional groove
	7	Functional cone
	8	Operating thread
	9	All others

Digit 3 — Internal shape, internal shape elements

	0	No hole, no breakthrough
Smooth or stepped to one end	1	No shape elements
	2	Thread
	3	Functional groove
Stepped on both ends	4	No shape elements
	5	Thread
	6	Functional groove
	7	Functional cone
	8	Operating thread
	9	All others

Digit 4 — Plane surface machining

0	No surface machining
1	Surface plane and/or curved in one direction, external
2	External plane surface related by graduation around a circle
3	External groove and/or slot
4	External spline (polygon)
5	External plane surface and/or slot, external spline
6	Internal plane surface and/or slot
7	Internal spline (polygon)
8	Internal and external polygon, groove and/or slot
9	All others

Digit 5 — Auxiliary holes and gear teeth

No gear teeth	0	No auxiliary hole
	1	Axial, not on pitch circle diameter
	2	Axial on pitch circle diameter
	3	Radial, not on pitch circle diameter
	4	Axial and/or radial and/or other direction
	5	Axial and/or radial on PCD and/or other directions
With gear teeth	6	Spur gear teeth
	7	Bevel gear teeth
	8	Other gear teeth
	9	All others

(Other Digit 2–5 categories apply to Digit 1 = 3–9)

Fig. 2-3. The Opitz Group Technology system.

that must be controlled, and the wider applicability of those parts, enable production and inventory planning and control to more effectively manage the acquisition and control of those components.

In engineering-oriented companies and those whose engineering departments exhibit a large measure of autonomy, there tends to be a proliferation of unique, although similar, parts. Engineers are taught to find the best solution and therefore are not motivated to seek out acceptable parts that are readily available. Enlightened engineering management will recognize the folly of this traditional outlook, and will include in the management systems, policies, and staffing of the engineering function a provision for identification of parts that can serve common usage and a reduction of unique parts where standard parts can be used. The engineering release process (quality/approval function, design review board) should be tasked to focus on this objective.

The second group of characters in the Opitz number, which is four digits long, holds dimensions, material, original shape of raw material, and accuracy classifications — all information that applies to the production process. The optional suffix is used to further define production information (operations, sequence) and can be unique to the company.

Using the second and optionally the third groupings, designers can consider how the part will be made, and can involve the production engineers more easily in the design stage. Production planning can also be a factor in how the part is to be handled, which plant resources will be called upon in the production process, and what is the most economical production method.

Computer Aided Engineering

The computer-stored geometric definition of the part can also be made available to printing and plotting devices, transferred to another system for use by other designers, and made accessible for other applications.

The two general classifications that describe the majority of the secondary uses of stored geometry are Computer Aided Engineering (CAE) applications and Computer Assisted Manufacturing (CAM) programs. CAE includes all design and development related analyses, while CAM includes only those activities that relate directly to support of the production process.

The most basic of CAE applications concern themselves with the testing of the physical properties of the object based on the mathematical description. Among the properties that can be tested are weight when made from various materials, center of gravity, how the object will react to stresses, and the heat transfer characteristics.

Chief among the listed items is the object's reaction to stresses. Since an object can be subjected to stresses of many kinds, at any point, and from many angles, this can be a very complex analysis. To simplify the task, and to make it feasible from a data processing perspective, the analysis programs will consider the stresses as applied to each of a number of individual small segments of the object. By breaking the large problem down into a number of small problems, the task is made manageable. The results of the individual analyses can then be displayed grouped together to allow assessment of the impact on the entire object. This approach is called Finite Element Analysis.

Stress analysis involves many complex calculations and therefore demands a lot of computing power to get the job done. Typically, the system used to develop the geometry is also the system best suited for analysis since both tasks are computationally demanding. The development of the workstation computer, which is optimized for computational speed, has been driven primarily by this kind of application.

In the workstation marketplace, the marketing people like to emphasize "MIPS" and "FLOPS" as measures of this processing power. MIPS stands for millions of instructions per second, which is an indication of how fast program steps can be processed. FLOPS are floating-point operations per second, related more to the efficient processing of standard mathematical functions contained within the program instructions.

In a continuing effort to stay ahead of the competition, computer hardware companies have developed a new technology called "Reduced Instruction Set Computing," or RISC. The idea here is that the processor can be made to work faster if there are fewer kinds of instructions that have to be interpreted and handled. By simplifying the tasks, more tasks can be handled per second given the same processor technology. With the introduction of RISC, the MIPS numbers soared.

There are some tradeoffs here, however, that can make the MIPS numbers misleading. If the instruction set is simplified, then more individual instructions will likely be needed to perform a given task. When comparing a "conventional" system at, let's say, 1 MIPS to a RISC system at 5 MIPS, don't expect a fivefold improvement in throughput. The RISC system will require more instructions for the same task and therefore will be less than five times faster overall.

While the concept of RISC is universal, the implementation is not. Various implementations of RISC have different instruction sets — some more simplified than others. The completeness of the instruction set will have an effect on the ultimate efficiency of the user programs. Also, compilers[1] are now available that can optimize the programs by reorganizing and rearranging the program steps for the most efficient execution by the system. MIPS and FLOPS do not translate directly into throughput. The only way to know for sure is to run a side-by-side test with your application or one which puts similar demands on the systems.

Other CAE tasks include analysis of heat transfer characteristics and testing of other physical properties of the design including expansion and contraction, weight and balance, sheer strength, and deformation characteristics.

Another area of CAE applications is kinematics, or simulated motion. If the design includes moving parts, the motion of those parts can be calculated and displayed on the screen as a "motion picture" to allow the designer to visually check fit, coverage, and alignment, and look for interference with other design features. The interference checking process can also be used to simulate assembly procedures for parts that do not normally move in use but must pass each other as the assembly is made.

Data Exchange

When the engineering tasks are performed on the same computer as the design process, the transfer of information from the design applications to the engineering programs is pretty simple . . . the design file is "saved" then retrieved by the analysis program. As long as the analysis program can interpret the information in the file, there is no problem.

Analysis programs that were written to work with a specific design package (or so adapted) are capable of using the data file description of the geometry as is, or contain an internal translator to convert the data to a usable format.

If the applications were not specifically designed to work together, then data exchange becomes a major concern. Since each vendor is free to define data files in any way, there are many incompatible formats in existence. There are three approaches to exchange of data between incompatible systems: the receiving program can be adapted

---

[1]A compiler is a system utility that translates computer programs, as written by the programmer in a higher-level language such as FORTRAN or C, into a form that is directly usable by the computer. The original program is human-readable, the compiled program or "object code" is not.

to interpret the format of the source program; a translator can be used to convert the format from one to the other; or a "neutral" format can be used with the source data translated to the neutral format, exchanged, then translated from the neutral format to the recipient system format.

While the situation is presented in the context of a design system providing data to an analysis program, the same considerations apply when transferring information from one design system to another or between analysis applications developed by different vendors. These situations occur more and more often as companies and their suppliers and customers install CAD/CAE/CAM systems.

The first solution relies on the vendor of the receiving system to provide a translation capability. If the source system is widely used, perhaps the vendor of the receiving system will be willing to make the investment. There is a limit, however, to the number of versions of a package that a vendor will be willing to support. The vendors must also worry about staying current with new versions of the source system and any impact updates may have on the data that are exchanged. If the receiving system is the more popular, the source vendor might be willing to offer a compatible program. Again, you are at the mercy of the vendor and subject to the realities of the marketplace.

The second solution—specific translation packages—offers an efficient conversion (only one translation required) and perhaps some independence from the application vendors. There are a number of companies that have made a successful business of translation packages, or you can write the package yourself. Direct translation works well if there are a limited number of systems or applications. But since a separate translator is required for each pair of applications, the number of translators can grow.

Take the example of a company that makes machine parts for the automotive industry. To accept design data directly from Ford, GM, and Chrysler, assuming that the vendor's design system happens to be compatible with one of the "big three," two translators would be needed. Add Navistar, American Honda, and a government contractor, and perhaps three more translators would be required. On top of this, there is still the problem of updating by either of the vendors (source or receiving).

Using a neutral format has the obvious advantage of being vendor-independent and, if the neutral format is relatively well accepted by the industry, major vendors will take the responsibility of providing translation capabilities to and from this neutral format and maintaining compatibility through the upgrade process.

Efforts to establish a neutral data exchange format for graphical data began around 1980 with the release of the Initial Graphics Exchange Specification (IGES). IGES was developed by a group of prominent CAD/CAM users and the U.S. Government, and was published by the National Bureau of Standards. Initially addressing only graphical information (between CAD systems), IGES is now in its fourth version, and includes constructive solid geometry and the data necessary for finite element analysis. It is currently the most widely accepted exchange format but has been criticized as deficient in handling other data such as bill-of-material information, and as inefficient in the use of file space. An IGES file can sometimes be four or five times as large as the original data file from the CAD system.

A more recent effort is the Product Data Exchange Standard currently under development by a consortium of eighteen companies. While it is still in the early stages of development, PDES is intended to offer a more comprehensive set of data in order to support more manufacturing information such as tolerances, features, materials, and finishes.

The International Organization for Standardization (ISO) has formed a subgroup to develop an international standard for graphics data exchange. The proposed standard, known as STEP (STandard for the Exchange of Product data), is based on PDES and will hopefully become the eventual worldwide standard.

As these standards become more accepted, most vendors will either modify their offerings to utilize the standard format or offer optional translation programs to convert to and from the standard. A major problem during the development stage (of the standard) is that the standard tends to change over time requiring the vendors to continually try to keep up. This explains a general reluctance among vendors to jump into standard formats early on.

CAD/CAE Today

As in all other areas, the plunging prices and rapidly increasing power of computer systems is changing the rules of the game. Whereas several years ago it took a mainframe system to support CAD and CAE applications, the development of the workstation computer has brought the entry price within the reach of small- and medium-sized companies. A mainframe system typically costs in excess of one-quarter-million dollars while workstation systems can be installed for under fifty-thousand.

Mainframes still have the advantage in storage capacity and the ability to support a number of users, but each new generation of workstations has pushed the performance capabilities of these systems to new levels. Often, today, workstations are networked together to allow convenient exchange of data files and sometimes are connected to mainframes which provide file management services.

CAD programs for personal computers (PC's) have made significant penetration into smaller companies and, perhaps surprisingly, are also found in many large companies as well. These systems, while not as powerful as workstation or mainframe implementations, are growing in sophistication and capability. There are now CAE applications, as well, for the PC market that take advantage of the newer, more powerful PC systems now reaching the market. As with the workstations, each new generation of PC's is considerably faster and more powerful than the last. The distinction between high-end PC's and the smaller workstations is now more of an intellectual exercise than a description of capabilities or limits.

Chapter 2 Review Questions

1. What are 2D, 2½D, and 3D graphical representations?

2. What 4 pieces of information are needed to add a feature to a CAD drawing?

3. What is the difference between boundary and constructive solid models?

4. What are primitives?

5. What are the groupings of characters in the Opitz system used for?

6. Should you buy a workstation based on MIPS?

7. What are the two current CAD data exchange standards, and how do they differ? What is the most likely future international standard?

8. Why don't all CAD and CAE vendors use the standard formats?

Since there is no strict definition of where Computer Aided Engineering leaves off and Computer Assisted Manufacturing (CAM) begins, you may see slightly different interpretations of which category includes certain applications. Some authorities consider CAM to include only machine programming and shop-floor applications, while others include some of the manufacturing engineering applications in this category. In this book, Computer Assisted Manufacturing deals with the use of systems to help plan the manufacturing process (Computer Assisted Process Planning or CAPP) and to develop automated machine programs used in the production process. CAE is limited to design and engineering applications. Computer usage on the plant floor is covered in Chapters 7 and 8.

3. Computer Assisted Manufacturing and Engineering Release

Process Planning

Process planning in conjunction with design is becoming more important because it saves time in the design and development process. Overlapping these two activities (design and process planning) is called Simultaneous Engineering or Concurrent Engineering, and it recognizes that markets are moving faster than ever before and that delays in the design process can sometimes cause a new product to completely miss its market opportunity. With rapid technological change a fact of life, the company that can engineer and produce a new product more quickly than the competition, and get that product to market first, can often enjoy a brief period of relatively high margin sales before competing products cause margins to erode.

An additional benefit of simultaneous engineering is that it often results in better designed products. By considering the production impacts of design features, the designer can come up with a product that is easier to make, can be made more efficiently (cost savings) and more reliably (fewer rejects, less rework), or better utilizes the facilities and capabilities of the production plant.

When the implementation of manufacturing cells and the use of automated material handling systems, particularly robots, is considered, process engineering during design becomes a necessity. Not only must the design be compatible with the processes and handling devices that are available, but the potential load on the facilities should be considered when developing the process definition.

Manufacturing cells differ from traditional plant layout in that dissimilar processes are often combined into a relatively independent unit that can perform a series of process steps in one relatively compact location. (Cells are discussed in more detail in Chapter 8.) This arrangement reduces the amount of handling and travel "mileage" that the production part(s) must go through in order to complete the process. Traditional plant layout combines like processes (mills in one area, drills all together, lathes grouped together) and has the work move throughout the plant to meet the process requirements.

Since cells are specialized for particular processes, it is important to consider the needs of the part and the cell(s) that are capable of satisfying those needs. Often there will be similar, though not identical, cells available that can perform the steps required for a new part. In addition to considering the tradeoffs between the capabilities of the different cells, it is useful to consider the current load, projected future load, and the impact of the additional load imposed by the new part on the candidate cells. Obviously this can only be done if production planning and scheduling becomes involved in this portion of the design process.

Retrieval Systems

Planning of the production process, the generation of the manufacturing instructions or routing, has traditionally been considered to be a manual activity. Since each part is different, and the facilities and capabilities of each plant also differ, it might seem that there is little opportunity for automation of this process. In fact, however, there are many ways that computers can participate in this process yielding not only time savings but also better routings and more consistency.

There are two general approaches to Computer Assisted Process Planning: retrieval systems, and a generative process. Retrieval systems, as the name implies, are based on the idea that there are existing process descriptions that are at least in part applicable to the current need, and if they can be retrieved conveniently, their availability can help avoid duplication of effort.

Many times, a newly designed part has some similarities to parts that have been designed and produced before. If this is so, then using the routing (process description) of the "old" part as a starting point for process definition of the new part on a "same-as-except" basis can yield considerable time and effort efficiencies. The challenges are to classify parts according to processing characteristics and be able to retrieve similar parts when needed. As shown in the previous chapter, group technology systems have been designed to include process characteristics in support of this objective.

In the Opitz group technology system, the second group of characters is allocated to general production characteristics. Specifically, the first character of the second group (6th digit of the code, since the first group is 5 characters long) is for dimensions, the second character is a material classification, the third is original shape of the raw material, and the fourth digit denotes accuracy requirements. In addition, the Opitz system includes an optional "secondary code" which can be used for more detailed production information such as process and sequence. The secondary code (4 characters) is available for individual company definition.

Once a part has been defined in the design process, it can be coded into the group technology classification system. Similar existing parts can now be identified by matching the classification code in total or in part. The process engineer can now use the routings for the existing part(s) as a starting point for developing the routing for the new part.

Another way that group technology can be applied to process definition is to develop and store "generic" process information (a general routing containing the common process steps) for each classification of part. Once the new part is coded, the generic routing for the class is retrieved and modified to fit the new part.

A variation of this approach uses a generic part characterization referred to as a "composite part." The primary processing requirements for the composite part (which contains the physical features that define the GT code group or part family) are included in a routing which is the basis for development of the specific routing for the new part.

Both approaches (retrieving similar parts or generic part/routing) assume a "same-as-except" approach where the differences between the retrieved routing and the new routing are identified and entered. Most production information management systems provide the ability to copy a routing from an existing part to a new one to support this approach.

If manufacturing cells are part of your production facility, the group technology coding scheme would be set up to recognize the cells and their capabilities. The use of cells does not necessarily mandate the use of group technology, nor does GT require that the plant be rearranged into a cellular approach. The two processes do, however, work well together.

Generative Process Planning

Generative process planning is the development of the manufacturing process definition by a computer system without copying all or a portion of the definition from an existing model. The generative process is based on a set of rules and an amount of knowledge that is built into the system program which allows the program to actually develop the process description in much the same way a human engineer would.

The program would go through a decision process that applies the rules and decisions according to the characteristics of the part's geometry, material, etc. Generative systems are relatively sophisticated applications, require significant processing power, and are not infallible. Practical systems are limited in the kinds of parts that can be handled and the complexity of the process that can be developed. The input to such a system would be a coded description of the characteristics of the part, not unlike the classification system discussed above for group technology coding.

A major concern with generative systems is the system's knowledge of *local* manufacturing facilities and capabilities. Since each manufacturing plant is a unique combination of machines, people, and skills, it is highly unlikely that a generative system can be developed that could produce a complete, flawless routing for a specific plant.

Regenerative systems must have a provision to input a description of the plant's capabilities, and a human engineer must be assigned to review and adjust as necessary the output of the system.

Nevertheless, regenerative process planning can provide significant time savings in the development of routings and will perform the task consistently and tirelessly. Since the same rules are applied to each part, the resulting routings will be much more consistent than those developed by a number of different engineers.

PLC Program Development

In Chapter 7 is a discussion of the application of Programmable Logic Controllers (PLC) to the task of directing machine movement during the production process. Among other applications, PLC's are used to manage the moving parts of individual production machines:

the feed and positioning mechanisms, spindle speeds, and tool selection carousels of mills, lathes, drills, etc., during the production process. This is (generically) Numerical Control (or NC).

The PLC's must be programmed. The program will define the sequence of events, the speeds and durations, and the positioning information necessary to direct the machine in the performance of the task. While machine-control PLC's can be programmed right at the machine through key-pad entry and/or through recording techniques, it is also possible to generate the PLC program in the CAM system through analysis of the CAD-defined geometry.

As with CAE applications (Chapter 2), the stored mathematical description of the part geometry is made available to the NC programming system. This could involve a physical transfer from one system (computer) to another and/or one or more translations to make the data usable to the program development application.

As discussed in the previous chapter, the IGES standard was originally designed for CAD-to-CAD data transfer, but has been expanded to the point where sufficient information can be transferred to accommodate the needs of most CAM systems and functions. The emerging PDES and STEP standards are intended to be broad enough to serve existing CAD/CAE/CAM requirements, including NC programming, as well as new applications as they develop.

The NC programming application will calculate the motion and control parameters (tool path) for the production machine without human intervention. The CAD system display can now be used to "play" the tool path, in motion, and from different viewing perspectives. Once the path is visually verified, the software will generate the NC program using a postprocessor that can translate the motion into commands for the specific machine that will be used. Numerical Control PLC's from different vendors may use a different command syntax, therefore, the postprocessor must be specific to the production machine.

Once the NC program is complete, it is transferred to the PLC either through a local area network connection or via tape. Once an actual test part is run with the new program, slight modification to the program may be required. These adjustments can be entered at the operator station of the PLC and the program corrections stored at the machine. If the program is to be used on other machines or is to be used for any other purpose, you must ensure that a copy of the corrected program is returned to the central storage facility or computer that will use it. Be sure that an obsolete copy of the program cannot be mistaken for the corrected version.

PLC program generation is not limited to the traditional machining processes. In fact, the effective design and implementation of flexible manufacturing cells and robotic material handling systems is greatly enhanced by the availability of programming systems for these applications.

Robot motion control systems have similar requirements and functions as machine controls, however, the three-dimensional movement "envelope" and simultaneous movement of several "joints" within this range makes robot programming a difficult task. Many robots come equipped with "recording" systems wherein the robot manipulator (hand) is moved to the desired position with manual commands. As each successive position is reached, the location is recorded and the controller writes the motion control program to duplicate the positioning.

With CAD-based program development tools, the motion of the robot can be defined on the CAD screen (and viewed in motion) and appropriate commands generated by the program. As in NC machine programming, the control program is processed and "downloaded" where it can be verified and possibly adjusted at the robot itself.

This capability is particularly useful when several machines are operating in concert and their motions must be coordinated. This is often true in a flexible manufacturing cell in which a robot is used to move the work pieces from one machine to the next. The graphics system can be used to simulate the motion of the devices and develop the coordinated programs to control them.

Other CAM Applications

As an extension or perhaps a subset of process planning, there are systems which will develop production standards (time standards) based on mathematical modeling techniques. Another variant uses empirical data (experience) analysis to develop new time standards based on past performance history.

Development of machinability information, speeds and feeds, is another application that can be included in a generative process planning application or could be a separate program to be used with or without automated process planning capabilities. This application would substitute for or supplement experience, rules of thumb, and tables which are the traditional sources of speed and feed information.

Innovations in cutting technology have combined special-purpose robot controls with advanced cutting techniques to produce automatic cutting tables for many applications. A typical large-scale cutter consists of a large table to hold the work piece over which is suspended the cutter (torch, laser, plasma cutter, water jet cutter, etc.) which is attached to motion controls. The work piece is positioned and the cutter does its job quickly, efficiently, and reliably as programmed. While one piece is being cut, another can be loaded onto a second work holder so that the work holders can be "swapped" and the cutter can immediately begin on the next piece. Associated with the new cutter technology is the ability and desire to arrange the pattern of cut to get the most usable pieces from each sheet of raw material. "Nesting" programs have been developed which optimize the combination of parts and their positions to find the most efficient layout (least waste).

Another emerging technique is that of applying tolerances to the CAD description for better process planning and to enhance fit analysis (interference checking). These tolerance limits can be displayed on the CAD screen, for visual verification, as well as included in calculations within the system.

Engineering Release

It is very important in many environments to get the design into production as quickly as possible. If production and purchasing activities for lower-level and long lead time items can be initiated before the final design of the product is complete, additional time can be saved and the product can make it to market that much sooner. There is a danger, however, in releasing a design or portion thereof to operations before the design is complete. The design is apt to change, which could obsolete parts and materials already on order, in process, or on hand, or could require parts to be reworked to comply with the new design. These are expensive, wasteful practices, and the decision to release before design completion should consider the tradeoffs between shortened overall lead time and the potential costs of design changes after work has started.

It is also conceivable (and actually quite common in some situations) to have engineering change activity going on for a product that is in active production. The risks here are the same as previously mentioned: obsolete parts, rework, and general disruption of the production process. Since concurrent engineering and production is often a fact of life, we must consider the ramifications and establish systems and procedures to deal with the challenge in the best way possible.

Recognizing the importance of engineering release control is nothing new, but considering and addressing its impact in the design and implementation of integrated information management systems is still rather a novel undertaking. From a systems perspective, coordination of engineering activities with production (both planning and operations) involves linkages between what are likely to be incompatible systems and the building of functional relationships between vastly different software systems, databases, and personnel procedures.

Parts List Information

The design and development process, among other things, will identify the materials and parts that make up the item being designed. These are typically listed on the drawing in standardized formats and should include a complete list of all parts and materials along with their specifications. In a CAD-generated drawing, these items will also be included, of course. Most CAD systems make provision for entering part identification information in such a way that it is identifiable as such and can be retrieved apart from the graphics information. Often, the part data are in a separate "layer" of the drawing file or are stored in a subfile (usually identified as "attributes" data) in a specified format.

Part information includes the identity of the material or component, the quantity required per unit of the item on the drawing, reference information such as the "bubble" number (location ID on the drawing), description, specifications, etc. Engineers are taught to identify each component in the parts list.

To be useful for production planning and control, level information is required that unfortunately is not always present. A parts list for a cassette tape, for example, might look like this:

Item: Cassette Tape

Qty	Part #	Description
2	8071	Outer Shell
1	6629	Pressure Pad
1	7724	Pressure Spring
2	9907	Tape Hub
300	T945	¼ in. Tape (Feet)
4	F329	Self-Tapping Screws

For production planning and control, this list is both incomplete and inadequate. Operations must know the relationships of the parts, how they go together, and perhaps other level(s) of detail. The bill of material for this same tape cassette might look like this:

Item: Cassette Tape

Level	Qty	Part #	Description
1	2	8071	Outer Shell
1	1	5297	Pressure Assembly
.2	1	6629	Pressure Pad
.2	1	7724	Pressure Spring
1	2	9907	Tape Hub
1	300	T945	¼ in. Tape (Feet)
.2	0.1	T900	¼ in. Bulk Tape (Rolls)
..3	0.01	T100	Bulk Tape (25″ Wide Roll)
1	4	F329	Self-Tapping Screws

This bill of material shows the indented relationships between the parts and also includes several entries that were not on the parts list. Item 5297, the pressure assembly, is called out on this bill to reflect that the two components are assembled together to make a new item which is later attached to the cassette shell. The drawing may or may not show this intermediate step; but for production purposes, if it exists as an identifiable item, it must be in the bill. This structuring indicates that the assembly is made, then set aside for use in the final assembly process.

The "Level" column indicates the relationships between items. All level 1 items are a part of the final assembly process. Level ".2" items are a part of the level 1 item that precedes them in the list. A level "..3" such as the bulk tape is a component of the preceding level ".2", which in this case is the result of slitting the 25 inch wide roll into ¼ inch bulk rolls.

The engineering definition (parts list) may not be directly usable for production for two reasons: levels or material details may be missing, and the level relationships (structure as built) may be incorrect or missing. In addition, engineering may not have access to the production database to validate whether part numbers are correct, or even whether they exist in the production system. Before any system-to-system bridges are built, these concerns must be addressed.

The best solution is a combination of education and involvement. The engineers must understand how the information will be used and how it must be presented to be useful. Engineering and production must establish a working relationship that allows them to help each other bridge the gap between departments.

On the data exchange side, the CAD system must first be capable of capturing the parts list information and making it accessible for extraction. This is relatively straightforward on most systems since the parts list information is specifically identified in the CAD file as "attributes" or a particular layer of the drawing. The internal file format, however, is unlikely to be usable as is. An extraction program, often a program added on at extra cost, is used to extract the data and format them to a specific (receiving) system's requirements or to a standard neutral format. Since IGES does not accommodate parts list data and PDES is not as yet a widely accepted standard, most often the format is proprietary to the CAD vendor and must be translated for use by the business system. Most vendors that are active in this arena have an internal standard. An example is IBM's Engineering/Manufacturing Interface format (EMI).

The part or bill-of-material information can then be passed to the production system either through a direct communications link (typically using a local area network) between systems or via magnetic tape or diskette in what is sometimes called "sneaker net." Whatever the medium, the same concerns arise at the receiving end as to format. There is often a need to translate the data again into a format acceptable to the receiving system.

Once transferred and translated, the application programs in the business system must be able to accept and use the data. Production control and planning systems must contain bill-of-material information. (Manufacturing Resource Planning systems are discussed in Chapter 5.) Often, these systems are designed for interactive entry and maintenance of the data files. Therefore, the bill-of-material processor sometimes must be modified to accept input from a data file (what data processing people call a "flat file").

The contents of the input file should be edited to ensure that the items identified in the bills of material are existing items in the business system. If not, the user must be prompted to add unidentified items before the bills can be accepted. Some business systems allow for this process through an engineering change control system that will accept nonstandard items into the change control system where they can be managed until the necessary identities are established in the main control system.

In the case of changes, as opposed to new products, the acceptance process would be responsible for identifying differences to be posted to the bill-of-material file as well as any new items that are not identified in the production system.

The engineering release control process should be in a separate management system from the production control and planning system but should be able to interact with it. As in the case of a new product with previously unidentified parts, or when a design is not finalized and "released" to production, a separate control system allows management of this information without confusion as to what is released and what is still in the engineering process.

When a new design or a pending change is still in process, there is justification for alerting the operations people that a change is being considered. With advance warning, particularly in the case of a change, it might be possible to avert a large "buy" or commitment of substantial resources to an item or process that might become obsolete in the near future. With the engineering release control system in communication with the business system, the initiation of a change request can be made to "flag" appropriate portions of the business system so that buyers, planners, production control, and others can be warned that they should check with engineering (or the engineering system) before proceeding with new orders or long-term commitments.

The above discussion holds true for process information as well as bills of material. The routing, whether developed with the assistance of a CAPP system or manually entered in the engineering system, can and should be electronically passed to the production side if possible to avoid rekeying and the opportunity for error that handling engenders.

When information is created in one system and passed to another where it is used and perhaps modified, it is important to "close the loop" with a feedback mechanism to the originating system. Engineering may pass a bill to production that production will refine for better manufacturability or to reflect limitations that engineering was not aware of. Under this scenario, the two systems would have different bills—an unacceptable situation.

If engineering is called upon to modify the product in question, it is likely to start modifying the original bill which is not the one being used in production and planning. When the modified "old" bill makes its way over to production, either the original changes will have to be overlayed on top of the modifications or the modifications will have to be applied to the production bill. Either way, it's more work for production and results in a high likelihood of considerable confusion and a great chance of error.

Companies on the leading edge of their technology or market will be in a constant state of product development and improvement. For these companies, control of the engineering change process and effective coordination of engineering with production and planning is essential.

To complete the change control process, there must be a review and approval function that includes planning, production, purchasing, marketing, and finance/administration functions as well as the engineers. A well-designed engineering change system will facilitate the review and approval process through electronic mail-type facilities and on-line approval and commenting capability. Once the change has been "released" to operations, updating of the business system should either be automatic (electronic) or carefully controlled to be sure that the change information is entered on a timely and accurate basis. Part of this process should be a review of newly designed unique parts to be sure that the requirement is truly unique and that an existing or "off-the-shelf" part cannot be used. Group Technology can be of great assistance in this effort.

Again, if changes are made to the proposed change between the time of design and acceptance into operations, be sure that there is a feedback mechanism to update the CAD or engineering files.

Finally, a history of engineering change activity should be retained in an easily accessible file cross-referenced by item (part number), by project, end-item affected, and group technology code. This history should contain not only completed projects but also those that were not approved or were cancelled, and the reasons why, so that the same false starts can be avoided in the future.

Chapter 3 Review Questions

1. What are the two methods of CAPP, and how do they differ?

2. How can Group Technology assist in CAPP?

3. Why can there be no "universal" generative CAPP system?

4. Should production planning be involved in process development? Why or why not?

5. How can we accommodate the different programming commands that are used by different NC vendors?

6. What are three methods for programming a robot? Which is best?

7. Can a computer-developed NC or robot program be used as is?

8. What are the steps in electronic transfer of bills-of-material information from CAD to the business system?

9. What are the possible impacts of not passing pending change information to production and planning?

The physical interconnection of computer systems, and the transfer of information over those connections, are governed by the characteristics of the connection or network and by the protocols that are being used. In addition to direct connection from one device to the other, there are a number of commercial network products available on the market. A few of these have become dominant in the manufacturing plant, while others are more widely used in business and office environments.

4. Networks and Protocols

A protocol is a formal arrangement specifying how communications will be carried out, including control signals, data formats, error checking procedures, and priorities. Many of the major computer vendors have developed protocols for use with their systems. There are several protocols that have become de facto standards by virtue of their widespread use in addition to several international standards in existence.

After presenting some communications basics, this chapter will outline the organization and functions of networks and protocols, and will discuss several of the predominant products and standards in use today.

Communications

In order to pass information or commands directly from one system to another, it is necessary to provide a connection over which the electrical signals will travel. In the simplest case, it would seem, a direct connection can be made with a wire from the output of one system to the input of another. While this might be theoretically possible, it is only practical if the device connected to the computer acts exactly like a terminal with all of the signals, voltage levels, and controls that terminal status requires. When a device behaves this way, it is called terminal emulation, and the connection is governed by the rules imposed on terminals for that system. A more generalized approach will use a standard connection and follow some standard arrangement of controls and data organization.

Connections can be either serial or parallel. A serial connection uses one wire to carry the signal. Other wires are used for control signals. One control signal might signify "I'm ready to send a message." Another might be "I'm ready to receive a message," and so forth. Such an arrangement is standard RS-232-C whose connections are illustrated in Fig. 4-1.

Serial connections use relatively few wires and are generally useful over relatively long distances. Effective communications distances can be extended indefinitely by using modulator/demodulator (MODEM) devices to convert a digital signal to a series of audio tones that can be transmitted over ordinary telephone networks. RS-232 communications is limited to, at best, about 19,000 bytes per second (called baud) and, using ordinary voice circuits, often less than that. Virtually all PC's come equipped with at least one serial "port" and others can be added inexpensively. Serial connection is point-to-point, that is, one device connected to one port.

Parallel communications transmits an entire character (byte) of information at once by using eight separate wires, one for each bit. Parallel circuits, therefore, need eight signal wires instead of one. A standard arrangement for parallel usage is IEEE-488, sometimes called General Purpose Interface Bus (GPIB), which is used primarily for laboratory instruments.

While parallel can be faster than serial (up to one-million bytes (eight-million bits) per second), and up to 15 devices can be connected on one IEEE-488 circuit, the devices can be no more than 2 meters

Pin	Signal	I/O	Description
1	DCD	I	Data carrier detect
2	RD	I	Serial receive data
3	SD	O	Serial transmit data
4	DTR	O	Data terminal ready
5	GND		Signal ground (0 V)
6	DSR	I	Data set ready
7	RTS	O	Request to send
8	CTS	I	Clear to send
9	RI	I	Ring indicator

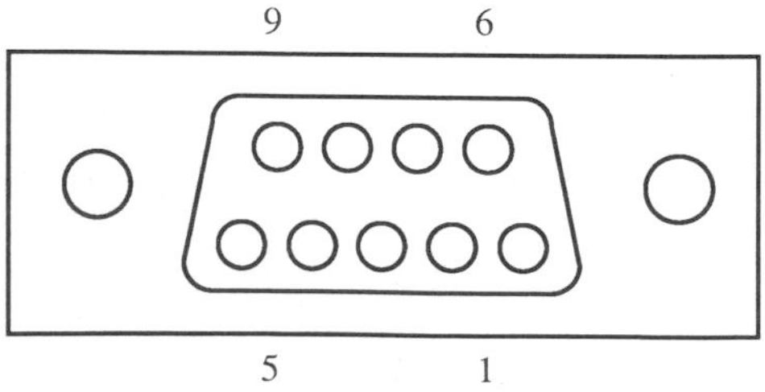

Fig. 4-1. RS-232-C pin (wire) assignment.

apart and the entire circuit can be no more than 20 meters long. Expander products are available, however, and IEEE-488 ports are off-the-shelf products for PC's. Obviously, parallel signals cannot be sent over telephone lines.

Networks

The idea of a local communications system that overcomes the limitations of ordinary serial and parallel communications is a natural outgrowth of the proliferation of PLC's, PC's, and other automated devices. It is very desirable to be able to connect many devices to a high-speed circuit and to be able to expand the arrangement as needed. The Local Area Network (LAN) is the answer to these needs.

A LAN is like a privately owned telephone system but with advanced capabilities specifically designed to handle data. LAN's are available from a number of independent vendors, and each has its own arrangements of wiring, transmission method, speed, controls, etc.

The arrangement of the wiring connections of a LAN is called the topology. There are three primary arrangements: Star, Ring, and Bus as illustrated in Fig. 4-2.

The type of wiring used is at least in part dependent on the type of signal that will be carried. There are two categories of signal: baseband and broadband, which differ in bandwidth. Bandwidth is a measure of the high-frequency limit of the system. When we mentioned that RS-232 had a practical limit of about 19K baud, that is a statement of its bandwidth and places it in baseband. The physics of the system limits its capacity to no more than 19,000 bytes of information per second. A system that modulates the data onto a radio-frequency carrier (a much higher frequency than the audio range carrier in baseband) can accommodate much higher data rates and is called broadband.

Baseband transmission is inherently less expensive and can use ordinary wires. In order to minimize the effect of electromagnetic

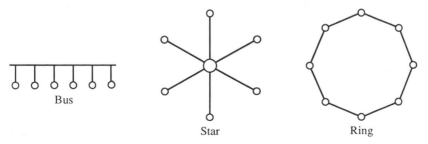

Bus   Star   Ring

Fig. 4-2. LAN topologies: Bus, Star, Ring.

interference, the wires are twisted so that each wire in a two-wire set will receive the same amount of induced signal from the interference and the interference can be filtered out and ignored. This is especially important in an industrial setting where electrical noise is endemic.

Broadband requires a type of shielded cable called Coaxial Cable or simply "Coax" (pronounced Koe-Ax). Coax has a center conductor surrounded by an insulator which is surrounded by a foil or mesh ground conductor and an outer sheath—the same kind of cable used in cable TV systems. Coax is more expensive than twisted-pair cabling and is less flexible (can't be bent as sharply), but one cable can carry many signals simultaneously (by using different carrier frequencies) including data, voice, and video. Coax is more resistant to electrical noise than twisted-pair because of its extensive shielding. Baseband signals can also be carried on coaxial cable.

The latest development in cabling is the use of fiber optics. Optical cable is the most expensive to install but is small in diameter, very flexible, and is impervious to electrical interference. Fiber optic cables have very high bandwidth.

Each commercial network system specifies the topology and wiring requirements. Ethernet, a widely used plant network, and DECnet, a similar system, both use baseband on coaxial or optical cable. Manufacturing Automation Protocol (MAP), a standard networking strategy developed by a group of users led by the automotive industry, is a broadband system using either coax or optical cables.

One of the primary characteristics of a network is the method used to control access to the system and how the message is passed. Ethernet uses a system called Carrier Sense Multiple Access with Collision Detection (CSMA/CD). In this system, any user (node) can send a message at any time. When the message is received, it is checked for interference (collision detection); if another message happened to be on the network at the same time and there was interference, then both messages would have to be resent. CSMA/CD systems are called nondeterministic. This means that it is impossible to predict, with any certainty, how quickly a message can be successfully transferred. As traffic on the network increases, so does the probability of collision, thereby increasing the time needed for (possibly multiple) retransmissions.

Another transmission arrangement is token-passing. This works like a railroad on which an engine is continuously making the circuit. When a node has a message to send, it simply attaches it to the engine the next time it passes. Token-passing is deterministic, which means that the time to complete a message transfer can be determined quite accurately and does not increase as traffic on the network increases. MAP is specified as a broadband token-bus network, meaning token-passing is used on a bus topology.

The Institute of Electrical and Electronics Engineers (IEEE) has issued standards for Ethernet (IEEE-802.3), Token Ring (IEEE-802.5), and broadband networks (IEEE-802.4).

Protocols

Since the late 1970's when the introduction of distributed processing ushered in a vastly increased need to interconnect computers, there has been a continuing campaign to define and implement communications methods. Each vendor has developed communications facilities that allow like systems to interact and, in most cases, to allow dissimilar systems within the vendor's product line to work together. Before the development of widely accepted standards, major vendors tended to set the tone by virtue of their presence in the market. In office systems, for example, IBM methods and protocols became a de facto standard, and many vendors developed communications systems that were compatible with IBM's System Network Architecture (SNA). In the plant, Digital Equipment Corporation set the pace, and many vendors can communicate with DECnet.

A protocol is a statement of the operating conditions that must be adhered to in order to communicate on the network. Generally, a protocol does not concern itself with the physical connection so much as control signals, data formats, addressing and routing of information, recognition of participants, security, connection and disconnection procedures, error detection and correction, and priorities of messages. A protocol might impose conditions on the physical connection but that is not its major concern.

In our overview of protocols, we'll start with the Open System Interconnect (OSI) standard which is the most widely recognized international standard, although not necessarily the most widely used format. OSI is open, which means that it is not a proprietary product of any one vendor and is available for use by any, and hopefully all, vendors. OSI is also very well defined and organized and presents a complete picture of the activities and concerns of a communication standard. After the overview of OSI, several other protocols will be reviewed and compared to the OSI standard.

Open Systems Interconnect

The International Organization for Standardization (ISO) has developed a data communications standard known as the Open Standards Interconnect (OSI) that includes a full definition of the physical and logical attributes of a data communications system. ISO has been a functioning international body since 1946 and has been involved in the development of OSI since 1977. Nearly 100 countries participate in this effort.

The OSI standard is defined in seven "layers" or segments that each address a portion of the interconnection process. This layered structure has allowed the standard to develop gradually, starting with basic physical connection and progressing upward in complexity to the application programs and how they are to interact. Layering also offers a measure of independence for each layer so that revisions in one layer brought about by technological change need not affect other layers of the standard. There is necessarily some overlap, however, in a layered structure, so OSI is not the most efficient approach that could be devised.

Efficiency, however, is never the primary goal of any neutral standard. Standards must be general enough to accommodate multiple vendors and multiple user tasks so there is bound to be flexibility built in that is not required by all users. This relationship, in fact, is a major reason why there are, and will continue to be, proprietary communications systems. Because a special-purpose system is designed specifically for a limited task, any functions or features not required for that task, and their associated overhead, can be eliminated.

LAYER 1, PHYSICAL

The first layer of the OSI standard defines the physical connection including mechanical, electrical, functional, and procedural considerations. Layer 1 does not specify a single connection type, but imposes requirements on the connection that can be satisfied a number of ways. There are several standard communication methods that are widely used to satisfy the layer 1 requirements including serial RS-232 and CCITT[1] standards (X.25) as well as most proprietary network products that comply with IEEE standards for Ethernet (IEEE 802.3), Token Ring (802.5), and broadband (802.4). Fiber optic networks as defined by the ANSI standard for Fiber Distributed Data Interface also qualify, and ISDN[2] inclusion is under consideration.

Layer 1 supports the transmission of data bits and does not concern itself with the content or organization of the data.

---

[1]CCITT is the Consultative Committee for International Telephony and Telegraphy — an international standards body.

[2]Integrated Services Digital Network — an international telecommunications standard currently entering service.

LAYER 2, DATA LINK

Layer 2 deals with blocks of data called "frames." A frame consists of a number of data bits packaged with a header and trailer that define the beginning and end of the bundle. The OSI name for the frame bundle and the definition of the header and trailer contents is High-level Data Link Control (HDLC). Other similar standards are IBM's Synchronous Data Link Control (SDLC), the IEEE 802.3, 802.4, and 802.5 definitions, and the ANSI X.25 standard—all of which can be used with OSI.

Layer 2, then, allows communications of bundles of information from one node of the system to another with confirmation of receipt. Layer 2 allows communication with other nodes on the same network (physically connected) but does not support "passing" of data frames into other networks and through switches. Proprietary networks such as Ethernet and MAP address only layers 1 and 2. Network software systems and proprietary communications products implement the functions contained in higher layers.

LAYER 3, NETWORK

This layer is responsible for routing the data to the correct recipient. The data themselves are divided into "packets," and a header is added to the data specifying the destination.

Remember that this description is working from the bottom up, whereas the activity on the transmit side is from top down. As data are passed down from the application, they pass down through layer 3 before reaching layer 2, therefore, the data are divided into packets first, then the header with the address information is added to the packet. This combined packet is passed to layer two where it may be further repackaged into the appropriate frames with their headers and trailers to assure proper transmission over the layer 1 physical connections. On the receive side, it works up through the layers, stripping off the layer 2 header and trailer, unbundling if necessary, and removing the layer 3 packet header.

Layer 3 specifies the requirements of the transmission network to route, verify, and reconstruct the message. Actual implementation of the requirements is not of concern to the user or the standard. Switching, routing, and message control are the responsibility of the communications network manager.

LAYER 4, TRANSPORT

Layer 4 manages the connections between the two user systems. This layer controls the establishment and termination of contact and the handling of the data packets. Layer 4 is concerned with end-to-end results and ignores any manipulation that occurs at lower layers. A typical layer 4 exchange would verify that a complete message was received and was received only once. This layer acts as the interface between the transport services which are provided in layers 1 to 3 and user applications in higher layers.

LAYER 5, SESSION

This layer controls the interaction of user systems. The establishment of intercommunication, its termination, and maintenance of the session are included. The exact nature of the session is determined by the systems, not the protocol. If two dissimilar systems are to interact, ground rules for the interaction and exchange of data must be established and will follow the parameters determined by one of the systems or perhaps a neutral format. Session parameters are included in the application standards for electronic mail, file transfer, terminal emulation, and program-to-program communication.

LAYER 6, PRESENTATION

This layer is concerned with the format and appearance of the data on the receiving end. Since there are many different ways to present information on a screen or in a file, the definition contained in layer 6 will ensure that the receiving system gets the data in the way it is intended and is useful.

Screen presentation is an area that has received a lot of attention lately with new releases of DECwindows, Hewlett-Packard's New Wave, IBM's Presentation Manager, and Microsoft Windows as well as the Open Software Foundation's OSF/Motif system. The majority of presentation systems are based on M.I.T.'s X-window format — some more closely than others.

LAYER 7, APPLICATION

The final and highest-level layer manages the interaction between the application itself and the communications system. This includes common support activities that allow access to data, manage security, identification of the users, sequencing of the interchanges, etc.

SUMMARY — FLOW OF ACTIVITIES

A user's application would interact with layer 7 which accepts requests for service directly from the user. Layer 7 works through layer 6 to define the presentation format, adding this information to the message. Layer 5 manages the session, adding session data to what it received and passing the whole lot to layer 4 which adds end-to-end verification controls. Layer 3 breaks the message up into packets, adds the "address" and routing information to the package, then passes the packets to layer 2 which adds "frame" headers and trailers. The resulting frames are sent over the layer 1 physical connections.

On the receiving end, layer 1 passes the bit stream to layer 2 which interprets the headers and trailers and verifies that communication was complete at the frame level, strips the frame information and passes the packets to layer 3. The packet address and routing information is interpreted and removed, then the information packets are relayed to layer 4. Here the end-to-end verification is completed (complete message, no duplication), the layer 4 information is deleted, and the rest moves to layer 5. Session controls are used and removed. Layer 6 interprets the presentation controls and passes the data to layer 7 which interacts with the application on the receiving end.

As you can see, each layer "communicates" with its counterpart on the other end. In effect, each message passed is a combination of messages, each with an appropriate recipient on the other end that interprets only its own part of the entire exchange.

THE UPPER LAYERS: APPLICATION

Layers 5 through 7 are addressed by several general categories of applications which are described in the OSI standard as well as being addressed by CCITT and proprietary products. These applications include electronic mail, file transfer, program-to-program communications, and terminal emulation.

Electronic mail systems are prevalent at most large companies but proprietary systems from different vendors are, of course, incompatible. The development of the OSI standard has supported further standardization including the issue of standard X.400 by CCITT. Compliance with the X.400 protocol by major vendors of electronic mail systems enables companies to interconnect various systems within their own company, to easily set up E-mail systems with vendors and customers, and to use E-mail for international message traffic. Currently, IBM's Professional Office System (PROFS), DEC's All-In-1, and Wang Office can all exchange mail since they each support X.400.

A second phase of the E-mail standardization, not yet finalized, is called Office Document Architecture, and it extends the definitions within the E-mail arena to recognize compound documents combining text, graphics, image, and even voice.

File transfer is accomplished according to File Transfer Access Management (FTAM). This capability is very important in manufacturing because of the need to transfer files from CAD systems to

CAE and CAM, to pass engineering information to production and planning, vendors, etc. The automotive industry strongly supports the use of FTAM among suppliers that are recipients of orders and design information electronically.

Other Protocols

PROPRIETARY PROTOCOLS

IBM's Systems Network Architecture (SNA) was one of the first communications protocols developed in response to the need for communications standards as systems became more distributed in the late 1970's. Like OSI, SNA is a 7-layer structure addressing the whole spectrum of communications from the physical layer to applications. Because of IBM's dominance of the mainframe market as well as its strong presence in smaller systems, whatever IBM does tends to become a de facto standard of sorts. Many other computer vendors feel that it is in their best interest to be able to communicate with IBM. Thus, it is logical that a number of vendors either have adopted SNA as a communications strategy or have built interconnections to SNA from their own protocols.

Digital Equipment has its Digital Network Architecture implemented in DECnet products. DEC also has a VMS/SNA software offering that supports connection of a VAX running the VMS operating system to an SNA network. Other vendors have similar capabilities.

TCP/IP

Transmission Control Protocol/Internet Protocol (TCP/IP) has its roots in the late 1960's in the U.S. Government's Advanced Research Projects Agency (ARPA). An alliance of government research organizations and educational institutions developed a network of interconnected computers called ARPANET to be able to more easily communicate and exchange information. The ARPANET project resulted in the development of protocols that were adopted as Department of Defense standards in the 1980's and gained widespread use outside of the defense arena as well.

TCP/IP, in fact, is the most widely used network protocol today and is expected to be a major player until the OSI standard eventually takes over sometime in the mid-to-late 1990's by most estimates. TCP/IP supporters speculate that even widespread acceptance of OSI will not completely eliminate TCP/IP.

TCP addresses the transport portion (layer 4 in OSI) for end-to-end communications and IP is the network (layer 3) specification. The majority of TCP/IP installations are on Ethernet (85%). TCP/IP itself does not address the applications side directly (layers 5–7) but is coupled with application protocols that specify how these requirements are met.

Electronic mail is covered by Simple Mail Transfer Protocol (SMTP). SMTP, however, is limited to unformatted text files, whereas X.400 will handle document format codes as well as compound documents. TCP/IP's file transfer protocol (FTP) is likewise limited compared to FTAM in that only complete files can be sent.

Despite its limitations, TCP/IP was available when OSI was just an idea. The large installed base and wide variety of equipment and software available will ensure that TCP/IP will continue to enjoy considerable popularity for some time to come. The TCP/IP management organization also accepts comments and suggestions from the user community and upgrades the standard to support the most needed features so TCP/IP continues to evolve.

Multiple Networks

It is not uncommon to find a number of separate networks within an organization. There might be several in the plant to support different production departments or focused factories, another in the front office, and yet another in engineering, for example. Not all of these networks or protocols are apt to be of the same type. There might be TCP/IP on Ethernet in the plant, SNA is found in the IBM office world, and perhaps the engineers are using a group of workstations on Token Ring.

Connections between like networks are relatively easy. They can be accomplished with either a repeater or a bridge. A repeater simply relays all signals that it encounters. If connected between two LAN's, all traffic is passed whether it belongs on the other LAN or not. Repeaters are inexpensive but nonselective.

A bridge will interpret the destination address (layer 3 of OSI) and only transfer a frame that must be passed to the other LAN. Bridges are not as fast as repeaters, but pass less traffic (only that which needs to be passed) so throughput may be as good or better. A bridge is, of course, more expensive than a repeater, on average, four to five times the cost.

Finally, if the networks to be connected are not using the same physical arrangements and/or the same protocols, a converting function as well as the bridging function is required. Devices that perform in this arena are called gateways. A gateway will provide a physical link using hardware adapters and also a software translation function to convert the protocol. Prices and capabilities for gateways vary widely because of the wide variety of tasks that a gateway may be asked to perform.

It is possible to establish a layered network arrangement with a "backbone" network to which are attached individual devices as well as subnets using bridges and gateways. The backbone network would not necessarily have to know the addresses of each node of the subnets; there are addressing protocols that would enable the entire subnet to be identified to the backbone network with the subnet handling the individual addressing among its nodes.

Beyond LAN's

Data communications often extends beyond the local area into the other companies (subsidiaries, partners, vendors, and customers) and to the rest of the world at large. Public networks are available to provide communications services as needed. Fortunately, there are international standard protocols without which large-scale communications would be impossible. An international standard that is in wide use is Integrated Services Digital Network (ISDN) which addresses not only data but also voice and video communications.

ISDN standards cover the first three layers as described in OSI (the network, not the user or application side), other CCITT standards and OSI itself govern the higher layer requirements. While ISDN compliance does not guarantee compatibility, it provides a framework and access to compatible worldwide networking services (the networks themselves are compatible with each other).

When communicating outside your local area, you can choose to set up private, direct connections (leased lines) or use public facilities. ISDN establishes hardware and network management standards for public networks. There are two levels of ISDN service; the basic level (Basic Rate Interface) includes two 64 kbps data channels in addition to a control channel. Primary Rate Interface includes 23 such data channels for a total capacity of 1.5 million bits per second.

Much "outside" communications today is handled through public Value Added Networks (VAN's). These provide such features as store-and-forward services using "mailboxes." VAN's are discussed more fully in Chapter 10, which addresses Electronic Data Interchange.

Chapter 4 Review Questions

1. What are the 7 layers of the OSI standard?

2. Which layers are normally addressed by a LAN product?

3. What kind of LAN does OSI specify?

4. What kind of LAN is Ethernet?

5. What is a frame? What is a packet?

6. Define CSMA/CD, and how does it differ from a token system?

7. What kind of network is specified by MAP?

The most prevalent management approach in manufacturing today is Manufacturing Resource Planning (MRP II), a philosophy which evolved from Material Requirements Planning (MRP) and is continuing to evolve through the incorporation of new ideas such as Just-In-Time. In this chapter we will review the development of the ideas behind MRP II and how the business and planning systems relate to activities and data in the plant floor and engineering areas of the company.

5.
Manufacturing Resource Planning

Material Requirements Planning is probably the most widely known of the manufacturing management techniques, thanks primarily to the efforts of Oliver Wight and others. Ollie and his colleagues made a career of promoting the virtues of MRP through seminars, video tapes, consulting, and books. Although he passed away in 1983, the Oliver Wight Company and associated firms continue to promote MRP and its derivatives including Manufacturing Resource Planning and the concepts of Just-In-Time.

In order to discuss MRP, it is helpful to define the philosophy and what we did before MRP. If you are already familiar with the "why's" and "wherefore's" of MRP, you could skip the next few pages — but why not go ahead and indulge me. It might be good for a reminder of how we got where we are.

MRP is a material acquisition philosophy which is designed to time the receipt of materials to projected needs. MRP relies on an anticipation of need such as a forecast or master schedule on which to base its calculations and recommendations.

Before or without MRP, material acquisition is typically based on keeping a certain level of inventory on hand. When the supply of an item drops below the predetermined minimum, more is ordered. This technique has been around as long as there has been manufacturing and is commonly known as "Order Point" (OP).

Picture the village blacksmith in colonial days. He might acquire a supply of iron from the regional iron works and draw from that supply as needed to support his production level. At some point in time, the supply will be used up and it will be time to hitch up the wagon for another trip to the works.

This scenario would be the most basic of order point systems in which the preset minimum level of inventory is zero. The obvious limitation here is that there is a period of nonproductive time when the smith is traveling to the works and bringing back the iron. If the blacksmith had an apprentice, he could send the apprentice out for the new supply of iron, and the blacksmith could at least continue to work on the last piece that he had withdrawn from stock until the job was completed. After that, it's time to sit under the spreading chestnut tree.

Now, if the blacksmith was smart enough and ambitious enough, he could send the apprentice out for a new supply before he ran out of iron completely. Then, hopefully, the new shipment would arrive before he ran out of work. We have now complicated things to the extent that the poor guy is forced to (gasp!) forecast his usage during the resupply lead time. He will now attempt to time the beginning of the replenishment activity (sending the apprentice on his way) to coincide with the time that the stock level has just enough left to keep him in business until the new supply arrives.

Let's not get that complicated just yet. We can benefit from a reserve supply without the complicated timing by simply having two wagons. We can start with both wagons full. When the first wagon is empty, send it down the road for refilling and start using the supply from the second wagon. In this way, as long as each wagon load holds at least enough to keep us busy until after the other wagon returns, we won't run out.

It should be pretty obvious that this system will result in a fairly large inventory investment. This technique is called the two-bin system—a form of simplified order point still in use in some places. The advantage is that it is a "no-brainer." No calculations are required and the reorder signal is crystal clear—an empty bin (or wagon). In fact, there is a modern technique based on the same principal (KAN-BAN) which is addressed later in this chapter.

Getting a little technical, the inventory levels in these two examples can be illustrated as shown on the graphs in Figs. 5-1 and 5-2.

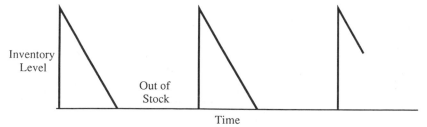

Fig. 5-1. Order point graph, OP = 0.

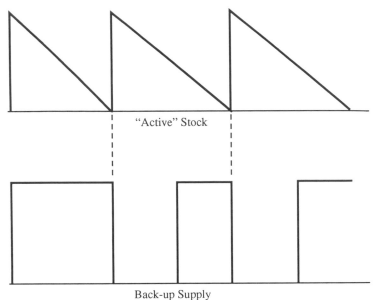

"Active" Stock

Back-up Supply
Fig. 5-2. Two-bin system.

Starting with a new supply, inventory is used until the supply is exhausted. With no back-up supply, as in Fig. 5-1, there is a period of time when the cupboard is bare. With the two-bin system, the back-up supply replaces the empty bin immediately so production can continue. As shown in Fig. 5-2, there is a period of time when the second bin is empty (resupply lead time). As long as the resupply is completed before the primary bin is empty, there is no problem.

With slight modification, the two-bin system can become the traditional order point system by simply replacing the second bin with a predetermined minimum stock level (Order Point). When the stock level reaches the order point, start replenishment activity (order some more). Now we are back to our blacksmith's nightmare — predicting (forecasting) the usage during resupply lead time. If usage is constant and resupply lead time is known, the graph can look like the one in Fig. 5-3. Timing is perfect in this illustration, i.e., the supply is used up exactly when the resupply arrives.

Of course, neither of the assumptions stated above is likely to be the case. Usage tends to vary from one day to the next, and lead times (due dates) may or may not be met. In order to compensate for some of this uncertainty, we will customarily add a "buffer" to the order point. Setting the order point at a higher level allows for some variation in usage above the norm after the replenishment activity is started. We don't have to worry about usage below the norm, because

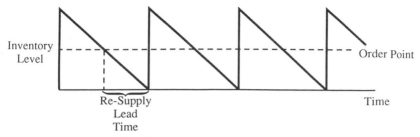

Fig. 5-3. Traditional order point without safety stock.

that will only result in having a little more inventory on hand when the resupply arrives. This causes us to start the next cycle at a higher level and will probably just postpone the next reorder.

The extra inventory represented by the higher order point is called safety stock and is used to help avoid running out if usage is higher than expected during resupply lead time (see Fig. 5-4). Since usage will vary, the logical way to deal with the usage estimate is to use an average. The one thing that is known about an average is that half of the time usage will be greater than the average, and half of the time usage will be less than the average. If we use average usage to set the order point, then half of the time we will run short before the resupply arrives, and half the time we will have enough (or more than enough) to cover the usage.

Since most people would not be content with a 50% "stock-out" rate, safety stock is used to improve that expectation. The real question is "How much safety stock is needed to cover any given service level (probability of having enough stock to cover demand)?" Fortunately, statistics can give us a scientific answer to that question. By studying the way the demand varies over time, statistical measures can be used to characterize the variations (standard deviation, for example) and a formula can be used to determine the safety stock level required for any given service level (50% to 99.99+%). This concept is illustrated in Fig. 5-5.

Why not cover all probabilities? The cost of coverage increases geometrically with desired service level as illustrated in Fig. 5-6.

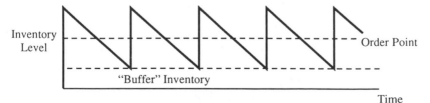

Fig. 5-4. Order point with safety stock.

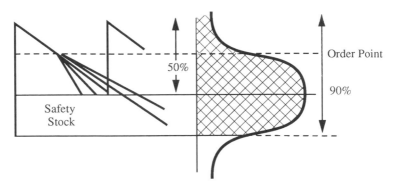

Fig. 5-5. Order point with normal curve and customer service examples.

However, it should be easier for a manager to decide on a service level and calculate the safety stock investment, rather than to guess at the safety stock level and not know what coverage has been obtained. In addition, the usage variation will be different for each item, therefore, a different safety stock quantity may be required to obtain the same service level on different items.

We've covered a lot of ground in a very little space. The net impact of all of this discussion is that order point is a common material management method that can be used in the absence of a planning system. Further, order point can be statistically monitored and adjusted to compensate for expected variations in demand during the period of risk, that is, during the resupply lead time. In the final analysis, though, order point assumes some level of risk of stock-outs because it is not possible to cover all possible variations in demand.

Order point has another major limitation. Order point bases all replenishment activity on remaining balance (or availability, which is defined later). Remaining, or on-hand, balance is the result of past

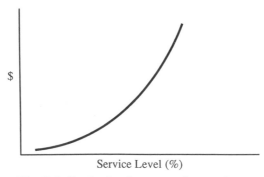

Fig. 5-6. Service level versus safety stock cost.

acquisition and use of the material. Order point systems don't anticipate demand. If usage rates change, they can only be covered by safety stock, luck, or by constant reevaluation and maintenance of the usage estimate (the order point).

There is a version of order point called time-phased order point that tries to compensate for this limitation. An estimate of need (quantity and date) is derived from sales or usage projections. Then a step-by-step simulation of usage and future receipts identifies required reorder activities. This process is used most often by distribution companies.

When order point is used for components (or raw materials), the shortage assumption (service level is never 100%) works very dramatically against you. Since service level is the probability that an item will be available in sufficient quantity when needed, these probabilities will multiply for all items needed at one time. If, for instance, a manufacturing order requires five different components, and each is managed (safety stock buffered) to a 90% service level, the probability of having all five available at any given time is $0.9 \times 0.9 \times 0.9 \times 0.9 \times 0.9 = 0.59$ or approximately 59% (a little better than half-and-half). Not very good odds.

Using order point for components also tends to result in unnecessarily high inventory levels despite the shortage assumption. Order point also doesn't anticipate changes in demand and will usually leave you with obsolete inventory when change eliminates the need for an item (unless manually consumed or the reorder recommendation is ignored).

Material Requirements Planning (MRP) takes a completely different approach to replenishment planning. MRP will only plan acquisition of an item if there is a need for it in the future. Further, the receipt of the replenishment stock is timed to coincide with the need, such that materials can be scheduled to arrive just-in-time.

In order for MRP to know what is needed, there are three major pieces of information required: a demand statement for end products, usually in the form of a forecast translated into a master production schedule (MPS) or build-plan; bill-of-material information to identify the components needed by the planned production process; and inventory information to calculate when existing stocks will run out creating a future shortage (net demand).

Once expected shortages are detailed, a lot-sizing or order-planning step may combine the needs into economic lot sizes. Then the start date for the acquisition activity is derived from the due date and the purchase or manufacturing lead time. The thing that makes MRP so powerful is that, if everything goes right, inventory levels can be reduced at the same time that availability is improved. This is a true "win-win" situation.

The problem, of course, is in the phrase "if everything goes right." "Everything," "always," and "never" are words that don't exist in the real world. The management challenge is to focus on the dependencies of MRP, understand their place in the scheme of things, and do everything we can to address the management realities involved.

Bills of Materials

The foremost dependency of MRP is on bill-of-materials data. Starting from the production schedule for end items, the first step in the MRP process is to identify the materials needed for that production order. The "Gross Requirements" are identified by multiplying the order quantity of the item to be produced by the quantity-per of the component. This represents the total quantity of the component needed to satisfy the production requirement. If the quantity-per is incorrect or the item is misidentified in the bill, the planning system will plan the acquisition of the wrong item (or the right item in the wrong quantity).

The quantity-per that is carried in the bill must be accurate *as used in production*. Remember that the purpose is to have the right materials at the right time. Frequently, actual production requirements will differ from the engineering bill of materials. The only way to validate and maintain bills is through accurate reporting of usage and a continuous program of follow-up and reconciliation.

Inventory Accuracy

After gross requirements are determined, the next step is to match the needs, day-by-day, to the expected inventory position on the need date. This requires that you know not only the current on-hand balance, but also what is expected to be used between now and the date of interest, plus any expected receipts during that time. This is the concept of availability and is the responsibility of the inventory system.

Inventory accounting systems handle transactions (reports of inventory movement) and keep track of the on-hand balance. An additional link in the information chain is the ability to track an allocation; that is, a demand for inventory which is needed in support of production or customer order activity, which (inventory) has not yet been issued from stock. An allocation is a reservation. It is not necessary to reserve particular items, merely to keep track of the total quantities required. For the expected receipts (on-order quantities and dates), the availability tracking system must be linked to manufacturing order and purchase tracking information to identify the expected future receipt of the item, the quantity, and date expected.

So, inventory accuracy involves much more than just an accurate on-hand balance. Order activity including due dates must be accurately maintained if the MRP system is expected to generate accurate plans.

MRP is designed to work as a "just-in-time" replenishment system in the strictest definition of the words. All else being equal, the plan generated by MRP will result in 100% availability of parts and materials with no inventories. The reason that it doesn't work that way in practice is that all else is never equal.

In order for MRP to be a zero inventory system, there can be no room for error in the operational dependencies. The bills of materials must be 100% accurate and inventory records would also have to be 100% accurate. In addition, all order activity would have to follow suit, including meeting all due dates. The easiest way to compensate for the unknowns and inaccuracies in your planning system is with extra inventory. If there are errors in the bills, such as the wrong item being listed or the quantity required set too low, then an extra quantity of the right parts could cover the problem. The same is true concerning accuracy in the inventory information. If there are extra stocks, we don't have to manage as closely. If the vendor is late, we might have some of these extra parts available that we can use.

In any effort to become more efficient and reduce inventories, you must be aware that you will be reducing or removing these buffers. Successful operation of your plant will be more dependent on the proper functioning of the information control system. Accurate and timely transaction-reporting procedures are required. Bills of material, inventory records, expected usage and receipt information, and lead time estimates must be highly accurate to be successful.

An honest and successful effort at MRP implementation is still unlikely to result in zero inventory. It will, however, improve availability, which is the primary benefit, and at the same time it will allow you to reduce your inventory "buffers" thus improving the effectiveness of your inventory investment.

Order Planning

Another reason MRP does not result in zero inventory is that the exact quantity needed is not the only consideration when deciding what quantity to make or buy. Even in the new world of just-in-time, the concept of an economic order quantity (EOQ) analysis is still valid. Tradeoffs between the fixed and variable costs of production and purchasing balanced against inventory carrying costs are as true today as they ever were.

In Fig. 5-7, the straight line ascending from the lower left represents the carrying cost for an item. It increases in proportion to the lot size because the more you make or buy at one time, the more you will carry in stock and for a longer time. The curved line descending from the upper left represents the fixed cost plus the per-unit variable cost to produce a part which decreases geometrically with larger lot sizes. The "U"-shaped curve at the top represents the total unit cost for the item (the sum of unit production cost plus carrying cost). It is obvious that the minimum overall cost is the low point of the upper curve.

When looking at this kind of analysis, be aware that most of the numbers used in this analysis are not really well known. For example, people are hard pressed to come up with definitive numbers for carrying costs (cost of money plus space plus personnel plus insurance plus risk of obsolescence plus damage plus whatever).

While the relationships remain true, the factors are changing. I believe that carrying costs are a lot higher than we have believed in the past. Most accountants will quote a figure of 40–50% (percentage of acquisition cost for the item) per year. EOQ analysis in the past often assumed a much smaller figure. The true figure (if there is one) might be even higher. At the same time, automation has reduced machine set-up and changeover requirements and increased manufacturing flexibility, thus reducing the fixed costs associated with production. Inventory reductions and other efficiencies also contribute to the lowering of the unit cost curve. The net effect of both of these trends is to drive the EOQ curve to the left, toward smaller economic lot sizes.

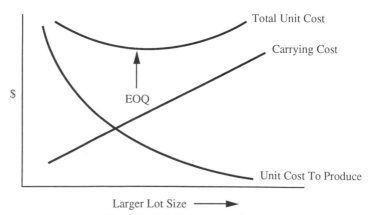

Fig. 5-7. Economic order quantity (EOQ).

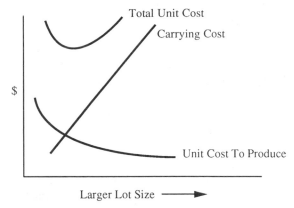

Fig. 5-8. EOQ with new assumptions.

In any case, economic order planning rules attempt to apply outside factors (such as transportation considerations, handling costs, packaging considerations, price breaks, etc.) to the required quantities needed to develop the "best" order sizes, which will often be higher than the exact quantity needed on any particular day.

Any order planning decisions that increase order sizes beyond the exact quantity needed to satisfy the requirements will result in the extra items going into inventory. The remaining quantity will be subtracted from the next requirement, possibly reducing the size of the next order or eliminating it altogether.

Lead Time Offset

The final step in the MRP process is to determine the start date for the replenishment activity based on the due date (need date) and the order size. If the lead time used by the system is longer than the actual lead time, the materials will arrive before they are needed and will build inventory. I call this management technique "just-in-case" rather than just-in-time. If planning lead times for purchased items are overly optimistic (plan lead time less than actual required lead time), orders will be released with too little lead time for normal accomplishment. In this case, either the parts will be late, disrupting the production schedule, or exceptional measures will be required to shorten the actual lead time (expediting).

For the release of planned manufacturing orders, the same considerations apply, except that the start date for the production order, rather than a purchase order, will be too early or too late. In either case, the production control and scheduling system will react to the improper release date with a higher priority for orders that start out late and a lower priority for early-release orders.

Master Schedule

In addition to the above, there must be a Master Production Schedule against which to plan the availability of parts. The master schedule must go out at least as far as the total lead time to acquire all of the parts and materials and to make the product (Cumulative Material Lead Time or CMLT). The level of finished goods inventory, as well as your on-time shipment performance, is a direct result of the relationship between the master production schedule and the rate of sales or shipment of the products you produce.

The Master Schedule is a production plan (start date, due date, and quantity for planned production orders) for the items that you sell. The Master Schedule is developed from customer orders, a forecast, and/or desired finished-goods inventory levels compared to expected availability (on-hand less allocations plus on-order quantities).

The most difficult aspect of Master Scheduling is using judgment to formulate a plan that is the best balance between expected demand and the capability to produce. A Master Schedule that overstates the ability to produce, for whatever reason (plant capacity, availability of materials, labor availability, skills, technology, funding), will not be accomplished. The most common mistake in Master Scheduling is not being realistic in defining production capabilities.

After the Plan

Once the plan has been developed, it must be put into practice. MRP will recommend the release of purchase and manufacturing orders, as well as make recommendations for the adjustment of existing activities to better meet the requirements of the plan. Systems must be in place to see that the tasks are accomplished in line with the plan. This is where MRP II takes over from MRP: MRP II includes the integration of the execution systems (production control and purchasing), customer service, and financial applications with the planning system. These applications, working together, form one coordinated system sharing a single database.

MRP's recommended orders are released to a production control and material management system, and these execution functions are responsible for tracking activity and making status information available back to the planning system. This is called "closing the loop" and represents the feedback-dependent nature of the MRP II process.

The shop floor must report activities to the business system to support order status tracking, efficiency and utilization analysis, work flow and prioritizing functions, and job cost accounting. The planning system, in turn, will use this information to update the plan and generate refined instructions (priorities) to pass back to the shop floor.

Plant Activities

The release of a manufacturing order will result in a priority being developed for that order that reflects its importance relative to other orders in the plant. This priority information links the order directly to the Master Schedule. As activities are reported to the order tracking system, the priority will be updated, thus providing the plant personnel with revised priorities as the work flows through the plant.

The information that is reported back to the system includes order and operation (activity step) identification, the worker and work center, how much time was applied to the job (could also include machine time as well as labor time), and the resulting pieces produced and pieces scrapped.

The state of the art for automation of the activity reporting is bar-code data collection equipment (covered in the next chapter). While bar-code systems reduce the amount of writing, and speed up the reporting process, the operator must still participate in the reporting process. It is technically possible to connect automated machines directly to the business system for automatic reporting of production rates and quantities, but there must be a mechanism to identify the item being produced, the order number, etc. Direct connection is not common at this time.

With the increased use of plant-floor communications networks (Local Area Networks or LAN's covered in previous chapters), there is some movement toward the distribution of manufacturing order information, priorities, and instructions through the network to plant-floor display terminals. This allows shop personnel to have access to up-to-the-minute priority information and job status. The more capable systems now coming into the market allow the merger of MRP II priorities with detailed manufacturing instructions and graphic images from the design and drawing systems. These terminals and networks support split-screen capabilities to show the instructions (text) and graphics at the same time and some allow basic manipulation of the graphic images (rotation, zoom) at the display terminal. In addition to more timely information distribution, such a facility eliminates paper (always a plus) and provides access to more information than would otherwise be distributed on a routine basis to the plant floor.

Not all industries manage production using work orders. In continuous production environments, where the same product is produced in high volume for an extended period of time, the production order might be inappropriate. These industries manage the day-to-day issue of materials, tracking of activities, and accounting requirements based on the expected production rate and the actual production quantity per day or shift.

Continuous production management systems tend to be less rigorous than order-based systems, but must serve the same purpose in terms of tracking and accounting for activity. Material control in continuous production often uses either a modified order-point philosophy or a KANBAN system.

KANBAN is a Japanese word meaning "card;" it was first used to describe a material management system developed by Toyota which used cards to trigger production of certain parts used on many of their vehicles. A supply of the item is made available to the using activity, with KANBAN cards attached to each piece. When a part is used, the card is placed in a defined location where its presence there signals the production of a replacement part. In this way, each time a part is used, its replacement is put into production.

KANBAN is basically a refinement of the two-bin order point system described earlier. The existence of the card in a certain place works just like the existence of the empty bin in the stock room. KANBAN has an advantage in that the number of cards can be controlled to designate desired inventory (or production) level. Removing cards from the system reduces the level, adding cards increases it.

KANBAN systems are useful in controlled circumstances but are not a substitute for other planning systems such as MRP. The KANBAN principle, in fact, has been successfully incorporated into many MRP II managed factories as an effective management tool for certain types of parts (typically, common-use high-volume components). Some newer continuous-production management applications that are a part of MRP II system include an electronic KANBAN capability that manages the issue and replenishment of components and materials based on production rate, daily schedule, etc. Many KANBAN applications do not actually use cards, but can emulate the capability using data in a computer or by watching the part bins, tray carriers, or storage racks.

Purchasing Management

A similar requirement exists on the purchase order side to manage and track activities. MRP recommends the release and reprioritization of purchase orders, and the purchasing management function keeps MRP informed of the latest order status. MRP uses this information to refine the plan and feed updated instructions back to purchasing. Purchasing is the link between *our* plans and the vendors' activities.

If we recognize a need for a purchased item early enough to provide proper lead time to the vendor, we can then measure vendor performance and use this information to manage the resource, select vendors who can deliver, and/or negotiate to improve performance. It is increasingly important to manage vendor performance as we try to operate more efficiently, be more flexible, and strive to reduce buffer inventories.

It is no longer appropriate to focus only on price when we deal with vendors. Because more efficient operation (reduction of buffers) makes us much more vulnerable, poor vendor performance such as late delivery or poor quality is more likely to interfere with or shut down our production lines. The primary focus, therefore, must be on delivery performance and quality with less emphasis on price. We should be willing to pay a little more for reliability and allow the good vendors to make an adequate profit so that they can stay in business and continue to serve us.

I'm not saying that price is not important. Of course it would be foolish to pay too much. But it can be much more costly to pay too little and not be able to get the parts when you need them to keep your production running.

Stated lead times should be what you can truly expect to happen, and you should make every effort to respect them. That is, if a vendor is committed to deliver in five work days, order at least five work days before you need the parts. If the order is placed within the promised lead time, you may be tempted to abuse the relationship you have built with that vendor, and the vendor may be tempted to make a promise that can't be kept.

Development of strong, cooperative relationships with your key vendors is sound policy. Treat them fairly and they will respond in kind. This is not to say that the relationships must be exclusive. While single-source arrangements may be developed, they are not required to apply this policy.

Some MRP II systems allow the listing of future requirements on the purchase order for the vendor's information and planning. The P.O. would include two or more sections: the firm orders; a second group which could represent an advance order, which is subject to change with perhaps a penalty clause; and a third section which is for planning purposes only. The relationship would be prenegotiated, of course. This advance information allows the vendor to be more precise in resource planning and to plan production based on knowledge of your needs, in order to better serve you.

Some customers will also negotiate with a vendor to "buy" a portion of the production capacity without committing to an exact product mix until the appropriate lead time. Many smaller businesses have found that, through the use of some of these techniques, they can actually get better service from their vendors than some of the vendors' other, sometimes much larger, customers.

Automation is moving into this area of the business as well. There is great interest in the emerging technology of Electronic Data Interchange (EDI). Some large buyers of goods and supplies, such as the auto companies and major chain stores, are equipping themselves to transmit orders to their vendors via computers. Through EDI, the vendors can send acknowledgements and shipping notices electronically to the customer, and follow up with a transmitted invoice. The loop can be completed with electronic funds transfer from the customer's bank to the vendor's account. EDI is in active use in the aforementioned industries and is spreading rapidly despite the lack of true standards. There are several interindustry and international groups currently working on standardization of EDI formats and protocols.

Just-In-Time (JIT)

Much has been written in the last ten years about Just-In-Time (JIT) as the key to how the Japanese have been able to dominate certain international markets, including automobiles and consumer electronics. There is probably no technique or idea in existence that is more misunderstood.

Perhaps it is the name that conjures up images of a vendor delivering a part directly into the hands of the waiting production worker who immediately attaches it to the assembly. While this can sometimes actually happen (or nearly so) in an aggressively pursued JIT program, it is not at all what JIT is all about.

JIT can best be summarized in two phrases: war on waste, and continuous improvement. The thrust of the JIT philosophy, and it is a philosophy not a piece of software or specific technique, is that there is waste in every manufacturing company, and management should focus on identifying the wasteful practices and take steps to reduce or eliminate the waste.

A useful definition of waste is "anything that adds cost without adding value to the product." The thing that probably comes immediately to mind when you hear this definition is inventory. It obviously adds cost but doesn't change the product at all. If inventory can be completely eliminated, then we have eliminated a major source of "waste." If you check back a few pages, you will discover that

MRP is really a zero inventory system but doesn't function that way in practice only because we can't control all of the factors (accuracy, on-time delivery) closely enough and have to consider other things such as the economics of lot sizing.

MRP, then, is not only compatible with JIT, but actually is a significant tool that can be used in the implementation of a JIT program. What better way is there to plan and execute production activities, identify areas for improvement, and measure results?

So, JIT is an attitude and an approach that strives for improvement and efficiency. The way it is practiced in Japan and in many other countries around the world, the JIT philosophy incorporates precise management of inventory, the production process itself, quality, and people.

In the production area, handling and lead times are the obvious areas of interest. Any time a work piece is moved, a cost is incurred with no change in the part itself, therefore, any reduction in handling is an improvement. Many companies realize significant savings by rearranging the layout of the plant to reduce the "mileage" that parts accrue moving between traditionally arranged departments. Often, dissimilar machines are grouped together so that parts can be completed in a single area of the shop. Other machine groupings are set up for other kinds of parts that use a different assortment of processes. The "work cells" or "Flexible Manufacturing Centers" are set up according to the processes needed by groups of similar parts. The general name for this approach is "Group Technology" (cf. Chapters 3 and 8).

The longer the lead time to produce a product, the more advance planning is needed, the higher the work-in-process investment, and the less flexibility there is to change production to meet changing market conditions. One of the benefits of the flexible manufacturing cell approach is usually a reduction in lead time. Less travel, and therefore less nonproductive shop time, serves to get the products through the shop more quickly.

Another consideration that can help in this area is a reduction in set-up or changeover time which also reduces nonproductive time for the machine or center. Automation plays a big role in reducing changeover, as addressed in Chapter 7.

The biggest consideration in production lead time is overall management of work flow through the plant. Studies have shown that, in a typical production environment, 70 to 80% of the total lead time is nonactive. The majority of the nonactive time is attributable to "queue" at the work centers. Queue is the waiting time between the arrival of a job at a work facility and the beginning of the actual productive time at that work center. In noncontinuous production, where work travels in lots or orders, each work center will have an average queue time, but each individual job will wait more or less than the average depending on the sequence in which the jobs are run. In a strictly first-come first-served arrangement, all jobs will have actual wait times somewhat close to the average. Prioritization and the resulting sequencing decisions will cause more variation in actual waiting times from job to job.

The key to reducing overall lead time is to carefully monitor and manage the throughput of the plant. Orders must be introduced to the plant floor at a rate that relates properly to the production rate. If more work is introduced than is completed, queue times will increase, average lead time will increase, and the work-in-process investment will increase. Reducing the input rate will not necessarily reduce the above factors, depending on the facilities needed by the jobs (product mix and resource requirements) and other management factors. The production control facilities of the MRP II system should be able to provide useful information for monitoring and controlling work flow and queue.

Other techniques that can have an impact on overall lead time include reengineering products for more efficient production, eliminating levels in the bills of material which reduces handling and intermediate level inventories, automating the production process, and changing to a make versus buy philosophy.

Quality control becomes much more important when striving for increased efficiency and reduction of inventories. With the elimination of buffers, we cannot afford to throw away parts or products because there will be no replacements available. The concern for quality — doing it right the first time — applies equally to the purchasing side and the production side. We must insist on 100% usable parts from the vendors, or else we must accept the need to buy more than the required quantity and the additional burden of separating out the good from the bad.

In our own production facility, we have to allow for additional materials to replace any part that doesn't make it through to the end product. To completely eliminate lower level inventories implies that there will be no losses from that level up to the end product.

The focus of quality programs has changed from quality control to process control. Quality control involves measuring the part or product, rejecting or reworking what is not usable, and keeping statistics to plan for the losses. Process control is a continuous measurement of vital parameters at each step of the production process and detecting problems immediately. The source of the problem can then be corrected before more bad parts are produced.

The personnel aspect of JIT is the full involvement of all employees in the improvement process. Each employee is a quality inspector, and any production worker can stop production if a problem is detected. The advice of production employees is sought in searching for improvements. Often the people on the line are in the best position to see what is going wrong or what could be done better, without regard to education level or prior experience. "Quality Circles" are often used to encourage this employee involvement in the improvement process.

Just-In-Time, therefore, is just a refinement of the basic ideas that were incorporated in MRP II and other fundamental management techniques and programs. JIT tends to include an emphasis on housekeeping. The idea is that in a clean environment, inefficiencies and quality problems will be more visible and therefore easier to detect. Finally, attitude is the key feature of JIT. Management must recognize that there *is* room for improvement, and it must be dedicated to searching out ways to do better.

Engineering Interface

To close the loop on the other side of our CIM triangle, there must be a coordination between the business planning and execution functions and the activities in the engineering area. As new products are developed and engineering changes formulated, it is important to keep production and planning informed to avoid unnecessary obsolescence and rework caused by "surprise" engineering releases. Also, the exchange of information from engineering to the plant floor (changes as well as machine control program distribution) should be timed in accordance with the current production plan.

As facilities become available which allow easier interconnection between dissimilar systems, it should become more feasible to develop linkages between engineering, production, and the business and planning systems. After the connections are made practical, procedures and software must be redesigned to take advantage of the availability of information from the other areas of the company.

Direct interconnection is only the path over which integration will travel. It is relatively easy for one system to access information that was generated in another area (assuming the physical connection has been made and translation/formatting software is available, if needed), but quite another to be able to use the information effectively. Another important consideration is: how good is the information being transferred, and how willing are the owners to share it with other users?

In the case of engineering information being released to the planning area, someone on the receiving end must want to see it and know how to interpret what he/she sees. Most effective would be a change in the business system software to recognize the effect of released changes and pending changes and warn the users accordingly.

Finally, close connection between engineering and production and planning areas will serve to help shorten the design-to-production lead time, thus making the company more responsive to changing market conditions and therefore more competitive. Free exchange of information also goes hand in hand with the concept of early manufacturing involvement (in design and engineering), which is a cornerstone of the CIM approach.

Chapter 5 Review Questions

1. What are three problems with order point?

2. What are the operational dependencies of MRP?

3. What does MRP provide to the shop floor?

4. What does the shop floor provide to the business/planning system?

5. Why doesn't MRP result in zero inventory?

6. What is KANBAN?

7. How does the implementation of MRP II and/or JIT affect vendor relations?

8. Define JIT.

The term "data collection" refers to automated or technologically assisted methods of gathering input for computers. In the context of CIM, data collection usually refers to the reporting of activity information from the production floor, and reports of material movement, which are fed into the business and planning system for production control and costing, inventory accounting, and payroll applications. An extended definition would include automation assistance in the gathering of measurements and process infor-

6. Data Collection

mation to support the quality function including quality control and process control applications. This chapter will focus on bar-code as the dominant vehicle for activity reporting and data acquisition capabilities as they relate to process control functions.

Activity Reporting

Manufacturing Resource Planning systems (Chapter 5) depend upon a timely and accurate flow of information from the operations areas (production, inventory, receiving) in order to perform their functions in inventory accounting and control, production control including scheduling and priority management, manufacturing cost collection, and payroll. The information to be reported falls into two general areas: material movement and production activity.

The information itself is rather simple. For inventory issues and receipts, the transactions include item identification, quantity, transaction type (issue to manufacturing, receipt from a purchase order, etc.), date, and perhaps the identity of the person reporting, and other identifying codes such as location, lot number, reason, reference, or comments.

For production activity, you must identify the task (usually order and operation), the person and/or machine, date, hours, quantity produced (good and scrap), and other codes such as whether it is set-up or run time, whether complete or not, and reference or reason codes.

Timely and accurate reporting is vital to the planning and control system. These reports serve as the "eyes and ears" of the system and are the only way through which the software can perform its tasks in these areas. In fact, failure to recognize the importance of accurate reporting, especially in the inventory area, is probably the single biggest factor in lack of success with MRP or MRP II.

MRP relies enormously on accurate inventory records and accurate bills of materials. The only way to get the record accuracy needed is to establish and maintain the disciplines necessary to ensure that all inventory movement is reported on a timely and accurate basis. What makes this goal especially difficult to achieve is that there are a lot of people involved, typically, and it is often difficult to properly motivate them all to this goal.

It's not that they don't care, but more a matter of priorities, in most cases. If a production worker finds that he needs more of a particular component for whatever reason, he should be free to get what he needs to complete the job. What is his motivation, however, to report this additional issue?

He may, in fact, be motivated *not* to report it. If the reporting will cause "pain," such as the supervisor coming down on him because of the additional usage (maybe he damaged the originals), he is actually motivated to *not* report. Even in the absence of a negative motivation, his focus is probably not on inventory accuracy. His goal is to complete the job without undue delay. He may have the best intentions to report the inventory issue "later," after the job is completed, but it is easy to forget "later."

The solution is not necessarily to lock the stockroom. If it is hard to get parts, people will find another way to get them. It is not unheard of that a tow-motor happens to go out of control and crash through the stockroom door just when there are some parts problems! It is also common for production people to squirrel away "left-over" parts against future shortages rather than return them and report the return as is required for accurate inventory records and validation of the bill of materials.

The purpose of this discussion is to point out that, no matter how sophisticated and easy the reporting vehicle, it won't work if the workers are not properly tuned into the need for timely and accurate reporting. If properly motivated, the ease that mechanized reporting can provide will help make the transactions more accurate and rush them on their way at the speed of light.

Reporting Methods

The most common reporting method is paper and pencil. It is obviously the least expensive system to set up, but is also the least efficient. In order to write down the report of activity, the worker is usually required to completely stop what he is doing, pick up a pencil, find the reporting ticket or time sheet, and record the appropriate information. The written report is then reviewed by a supervisor, and passed to a data entry person who does the key entry. An error report is produced from the transaction processing, then someone goes back to the originator to try to resolve any detected problems.

Not only is this process inefficient in terms of the amount of time and handling involved, it is also slow. In most situations where manual reporting methods are used, transactions are collected and entered no more frequently than once per shift. Entry and processing takes place after the originator has left for the day, and the error resolution process can't even start until the next day. The results of activity reports from shift one are usually not available until late in shift two, at best, and corrected entries aren't posted until well into shift one the next day, if that soon.

It is also difficult to accomplish error correction a full day after the event. The ability to recall details of an activity declines significantly with time, especially if a number of activities are reported in a single day or shift. Try a little experiment: try to recall exactly what you were doing yesterday morning at 10 o'clock. How long did it take to complete whatever it was you were doing? What other details can you reconstruct?

The idea of applying automation to the data collection task is not new, by any means. Data collection systems have been around for more than thirty years. What is new is the superior capabilities that can now be offered as a result of the advances in computer technology that have taken place in the last few years. The data collection systems of twenty or even ten years ago were batch oriented. The automation simplified the actual recording of the event, but the data remained in the collection system until a group of transactions could be moved to the business system for processing against the production and inventory tracking records. Transfer of a batch of transactions could be done several times each shift, if desired, and limited error checking could take place in the collection system; these are significant improvements, indeed.

Today's system can provide "on-line" posting and verification. As transactions are entered, the information is checked against the control system's files to immediately verify the validity of the transaction: is this a legitimate order and operation? Were we expecting these components to be used on this assembly? Was the quantity completed at this operation greater than or less than the quantity expected? A report of significant errors can be relayed *immediately* to the reporting station for correction on the spot, not the next day. The increased accuracy that can result from immediate feedback is phenomenal. In addition, the data are immediately available for system use for tracking activities, setting priorities, and reacting to problems.

While there have been many media and methods tried over the years, bar-coding has become by far the dominant form of automated data collection. Bar-code is relatively inexpensive to install, easy to use, quite adaptable for many applications, readily available from a number of vendors and suppliers, and extremely reliable.

The "runners-up" for the data collection media are magnetic strips and optical character recognition (OCR). Magnetic strips can contain more characters per inch than bar-code and can be erased and rewritten easily, however, they cannot be read by noncontact devices such as the laser scanners used with bar-code, and they are more expensive to print or include on documents and labels. While OCR can be printed as easily as bar-code and takes up less space, printing defects are far more likely to result in reading errors (error probability is hundreds of times higher than bar-code). In addition, OCR is more difficult to read since the scanner must be oriented more precisely to the line of characters and OCR is not amenable to high-speed, noncontact readers.

Bar-Code Basics

The concept of bar-code is quite simple. Numbers and sometimes letters and special characters are represented by "characters" made up of alternating bars (rectangles) and spaces whose width and/or spacing defines the characters. There are a number of formats of bar-code in existence, and more are being developed to satisfy increasingly more varied applications.

In manufacturing, the dominant format is "3 of 9," also called Code 39. Each Code 39 character is made up of five bars separated by four spaces. Of the five bars, two are wide and three are narrow. One of the four spaces is wide, the other three are narrow. Fig. 6-1 shows the letter "J" in Code 39. As you can see, it meets the criteria outlined above.

The tight design criteria are designed to limit the possibility of erroneous interpretation. It can be expected that, with any printing method, there is a possibility of smears, spots, voids, or other imperfections. If one or more bars is damaged in the printing process, it is highly unlikely that the resulting bar pattern will contain exactly two wide bars out of five with exactly three narrow spaces and one wide space. Error rates with Code 39 and fair-quality printing are measured on the order of one in one-million characters read. With

BARS   SPACES
00110   0010

Fig. 6-1. Code 39 character "J."

good print quality, only one substitution error is expected per thirty-million characters read.

The basic character set of Code 39 includes all numbers and letters plus "-" and "." and "space" (39 characters in all). The arrangement of bars and spaces offers forty possibilities; the fortieth character is a start-stop symbol, typically printed as a star or asterisk. A bar-code symbol consists of a minimum amount of unprinted space called a quiet zone, a start-stop character, the data characters, another start-stop character, and a following quiet zone. There is a small "intercharacter gap" or space between each character. This makes Code 39 a "discrete" bar-code — the opposite of continuous which has no gap between characters. See Fig. 6-2.

CODE 39

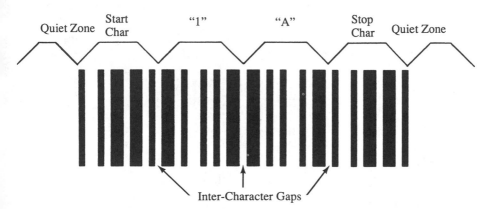

• 3 out of 9 elements in each character are wide, 6 are narrow ("Code 39")

• Characters are separated by intercharacter gaps

Fig. 6-2. A Code 39 bar-code symbol.

A bar-code symbol can be read (scanned) from either direction. The reader, the device that decodes the scanner signal, will recognize the direction from the start-stop character and decode accordingly. Since there are additional special characters that are sometimes required, four additional "control characters" have been defined (represented by $, /,+, and %) which can be used as modifiers for the basic characters. For example, an equal sign is coded as "%H," a # is "/C," and lower case characters are printed with the + symbol preceding the letter. These special symbols violate the two-wide-bars–one-wide-space rules, but are self-checking within the reader software definitions.

Most applications of bar-code include the use of a check-digit to further ensure data accuracy. Many schemes are available and their use is optional. Check-digits apply a mathematical formula to the digits of a number being printed and read, and through the use of an additional character, the check-digit, can detect a missing, extra, or incorrect character.

Code 39 is a variable length code, which allows the user to include as many characters as desired in one bar-code symbol. The practical limit is how long a symbol can be scanned in one sweep. There are standards for printing size and density (bar and space width), but there is a lot of flexibility once the minimum standards are met. Making the bars longer increases the ability to scan off-axis and still capture all of the bars and get a "good read" as shown in Fig. 6-3.

The scan need only cross the entire bar-code symbol at a reasonably constant speed. The reader receives an analog signal based on the differential reflectance of the bars and spaces. The reader must detect a change and determine the relative width of the symbol elements. With Code 39, the bars (and spaces) are either narrow or

Fig. 6-3. Bar-code symbol with sample scan lines.

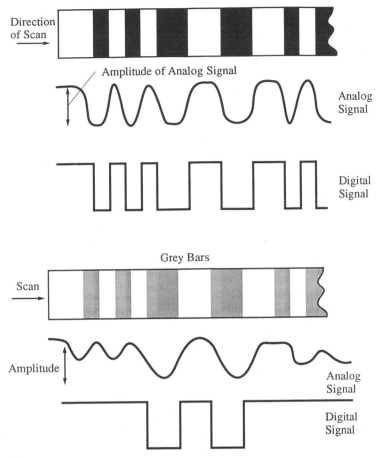

Fig. 6-4. Reading and decoding a bar-code symbol.

wide — it is a binary system. As shown in Fig. 6-4, if the contrast of the bar-code image is high, the conversion to a digital signal is easy. Scan speed doesn't matter, within limits, because relative width is used, not absolute width. This feature also allows the code to be printed in almost any size and still be readable as long as the scanner and reader can distinguish between bar and space, and narrow and wide.

With a low-contrast image, you can see that the decoding is much more difficult. If the scan is incorrectly decoded, chances are that the resulting pattern will not meet the two-wide–one-narrow criteria, and the reader would signal the user to scan again.

There are many kinds of scanners in use today. The two general categories are contact and noncontact. A contact scanner is typically a pen-like device that is manually "wanded" over the bar-code symbol image. Contact scanners contain a light source (either visible or infrared), some type of lens, and a detector, similar to the configuration in Fig. 6-5. Contact scanners can also be fixed-position devices with a slot through which the bar-coded document is drawn.

Noncontact scanners use a moving laser beam to illuminate the bar-code image with the detector capturing the reflectance from the laser beam. Noncontact scanners can read symbols up to fifteen feet away (depending on the model and the bar-code). They can be either hand-held (typically gun-shaped) for reading stationary targets, or mounted in a fixed position to read moving targets such as boxes passing by on a moving belt. Grocery store scanners are an example of the latter type.

Other Symbologies

Code 39 is the most widely used symbology for industrial use, and has been specified by the U.S. Government and by the major automakers as the required format for marking parts and supplies. Other symbologies (Fig. 6-6) have been specified for other uses.

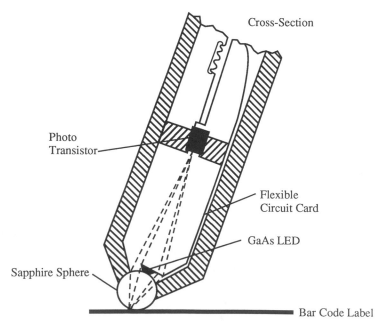

Fig. 6-5. Bar-code wand (contact) reader.

CODE 39

012345678905

UPC

0 12345 67890 5

INTERLEAVED 2 OF 5

012345678905

CODE 128

0123456789

CODE SET A

CODE 128

0123456789

CODE SET C

Fig. 6-6. Sample bar-codes.

UPC/EAN

Everyone is familiar with the Uniform Product Code symbols, which appear on virtually every item sold in a grocery store and on most other retail packaging as well. Designed to be relatively insensitive to variations in the printing process, the code symbol consists of two groups of five digits making up left and right half-symbols which can be scanned separately (by two different passes of the laser beam as the package is moved over the scanner). In addition to the five data characters, the left half contains a number system character and the right half a check-digit.

The relatively long bars accommodate fixed, noncontact scanners with the requirement that the symbol can be read no matter what the orientation of the package (as long as the symbol is facing the scanner). UPC is a numeric-only code and is formatted with a check-digit for further verification (the "5" in Fig. 6-7). The pairs of thin bars on each end and in the center are called "guard bars" which help the scanner determine if an entire half-symbol has been detected. There are no start-stop characters in UPC because each individual character has direction significance in addition to (parity) error correction.

The left five digits represent the manufacturer's number. These numbers are assigned by the Uniform Code Council which administers

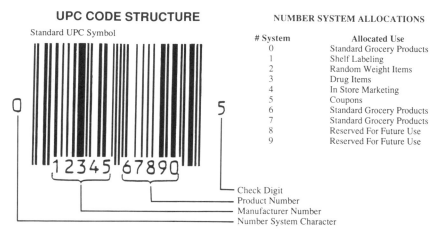

Fig. 6-7. UPC symbol.

the system in the U.S. The EAN system is the European Article Number, also known as World Product Code or International Article Number. This system uses the same symbology as UPC, and is said to be fully compatible.

INTERLEAVED 2 OF 5

The basic 2 of 5 code consists of five elements, two of which are wide. The resultant code is similar to Code 39 in many ways except that the spaces are of no significance and the code is limited to the 10 numeric digits plus separate start and stop characters.

In order to save space, an interleaved version of the 2 of 5 code was developed wherein the bars and the spaces were each encoded using the 2 of 5 scheme, thus "doubling up" the characters represented by a given length symbol. As shown in Fig. 6-8, the symbol represents the number "3852," wherein the 3 and the 5 are encoded in the bars, and 8 and 2 are in the spaces. The bars always represent the odd digits (first, third, etc.) and the spaces are the second, fourth, etc. Interleaved 2 of 5 can be used for numbers of any length (even number of digits) subject to the limitations of the scanning process.

Interleaved 2 of 5 has been adopted by the Uniform Code Council as the standard symbology for use on shipping containers (cartons). This symbology is relatively compact, and is quite forgiving of printing inconsistencies, but is less reliable (more subject to misreads) than some other symbologies.

CODE 128

Code 128 is a symbology that uses bars and spaces that can be up to four different widths. The widths are specified in multiples of the smallest used in a particular print size (called a module), i.e., if X is the width of the smallest bar or space, the other dimensions can be 2X, 3X, or 4X. This yields a compact coding system, but one that is more sensitive to printing variations. The character symbols are self-checking using parity on both the bars and spaces. Each symbol is 11 modules wide. (Fig. 6-9).

There are three separate start characters in Code 128, used to designate the use of Set A (Upper Case), Set B (Lower Case), or Set C (Numeric) symbols. A single bar-code symbol may contain more than one start character to signify a "shift" from one character set to another within the symbol.

Code 128 was developed as an improvement on UPC symbology and has gained widespread popularity in Canada. Knowledgeable bar-code industry observers predict increasing popularity for Code 128 because of its compact size and full alpha-numeric character set.

STACKED CODES

The newest development in bar-code symbologies is the stacked code. Several formats have been proposed in which a rectangular code symbol would contain several lines or rows of code characters (Fig. 6-10). Special readers are required for these new codes, which represent the densest codes presently available, although the printing requirements and scanning limitations are much more restrictive than more traditional symbologies.

INTERLEAVED 2 OF 5

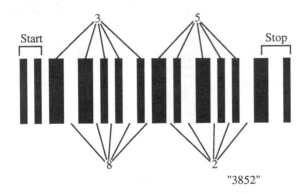

"3852"

Unique start and stop characters allow bi-directional reading

Alternate characters represented in bars and spaces ("interleaved")

Each character consists of 2 wide and 3 narrow elements ("2 of 5")

Fig. 6-8. Interleaved 2 of 5 symbol.

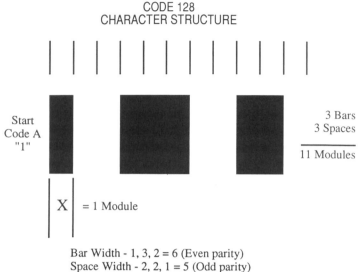

Fig. 6-9. A Code 128 character.

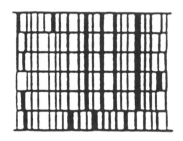

Code 49   Code 16K

Fig. 6-10. Stacked bar-codes.

SUMMARY

There are a number of bar-code symbologies to choose from, each with its own strengths and weaknesses. The safest path is to adopt the codes that have the backing of the industry in which the application belongs and/or are the subject of national or international standards. Trade magazines, seminars, and industry associations are readily available for interested parties. While scanners have limitations in terms of size, distance, and density of the bar-code that can be read, many readers are capable of decoding more than one bar-code symbology, sometimes identifying the type automatically.

Bar-Code in the Plant

The application that most commonly justifies the installation of a data collection system is the production tracking system. This function involves a lot of employees who generate a lot of transactions; the workers in the shop are least attuned to clerical functions; and the need for timely and accurate reporting is key to the usefulness of the application. Many companies, however, will take their first tentative steps into automated data collection for payroll reporting because it is more straightforward, relatively inexpensive, and easy to do.

Payroll, in most cases, only requires that the employee sign in and out once (each) per day. Reader devices are installed near the entrance and exit doors where they replace conventional time clocks. The system can provide various time and attendance reports on demand and save key entry of time cards. Readers at the doors can also serve double duty as part of an access control system. The reading of the badge along with a keyed (correct) password can generate a signal to operate an electric lock as well as provide a record of entry.

For production reporting, additional reader devices are needed in locations convenient to the production facilities. It is usually not necessary to have one reporting station per machine or work center, but enough devices, properly situated, so that the workers don't have to go too far out of their way to report.

Usually, two or three documents are required for such a system: the employee badge, a bar-coded process sheet (routing or traveller), and either a bar-coded sheet of transaction type codes and control codes (such as a "send" signal or "delete") or a reader device with keys for these codes. See Fig. 6-11.

Most production tracking systems expect a report of completed activity (or interim reports) including elapsed time and quantities produced. Data collection systems are oriented to event tracking. Starting work on a task is an event. There is a companion event, completing work, that completes the transaction. The system must be able to match the sign-on with the sign-off, compute the elapsed time, and format the transaction for the production system.

A typical "start work" transaction might include the following: enter or wand the "sign-on to task" transaction code, wand the order/operation identifier from the traveller, wand the employee badge, wand or key the "send" or transaction complete signal. The sign-off would include: the transaction code, badge, order/operation, quantity good, quantity scrap, complete or partial, and send. It may look a

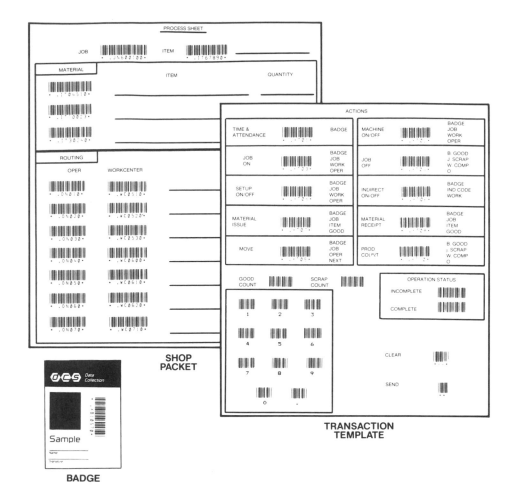

Fig. 6-11. Shop reporting documents.

little confusing in words, but it is very simple in practice and takes less time than writing it all down. In addition, the system does the math (elapsed time calculation) and can even adjust the times for normal shift start, lunch, shift end, and break times. Many systems provide a prompting capability to lead the worker through the data entry sequence.

The reader device will signal the user immediately, for each bar-code symbol, if the bar-code signal was read correctly, usually with a beep tone or a red or green light. Once the transaction is completed, the software will verify the entry. In data collection systems not directly attached to the production control system, the verification is only for completeness—were all of the data elements entered for a complete transaction of this type? For on-line systems, valid data can also be checked (valid order/operation, expected quantity, etc.) with a signal back to the user, within a few seconds, as to whether the entry was acceptable or should be repeated or corrected.

The process is similar for other transaction types such as material issues. The authorization document (pick list) would contain the bar-coded identification information (a bar-code symbol for each required component) and a different transaction type would be used. Most bar-code systems are flexible enough to include transactions that are not directly related to activity reporting such as: to report machine down, maintenance needed, send an inspector, etc.

EQUIPMENT

A data collection network (Fig. 6-12) can be controlled by the business system itself or by a dedicated computer which is in communication with the business system. The latter arrangement is more common. The typical business system is configured for interactive (human) users, and not necessarily well suited to tracking scores of shop floor entry devices.

The typical data collection system of today uses a PC as a controller. The controller manages the polling (watching the terminals for input and directing traffic) usually through one or more multiplexers which allow one communications port on the PC to serve a number of devices—usually 16, 32, or 64 devices per multiplexer.

The devices can be a mixture of wands, slot readers, and noncontact scanners. There are also hand-held devices that are equipped with battery power and memory so that they can be carried around for hours, collecting data, then plugged into the network when convenient to "dump" all of the collected transactions at once. Of course, without a direct connection, the only verification that is possible at the time of the scan is "good read" and possibly transaction completeness.

Bar-code readers can also be linked to the collection network through FM radio transmission. These RF (Radio Frequency) devices are especially useful for inventory transactions in large warehouses.

It is also possible to use bar-code devices directly attached to a system terminal as a direct substitute (supplement) for key entry. The reader is connected between the keyboard and the system. When the terminal is ready for, say, an item number, instead of pushing keys for entry, use the reader to scan the bar-code label on the item. The terminal doesn't know the difference between key strokes and bar-coded entries. This type of reader is called a "wedge."

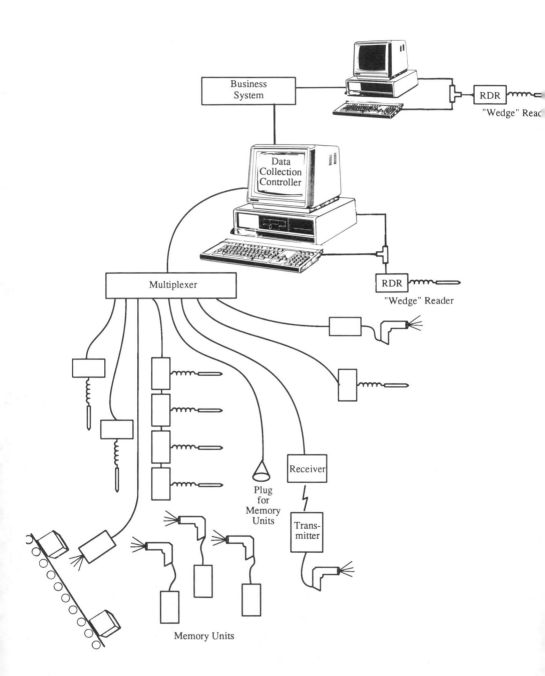

Fig. 6-12. An example of a data collection network.

Voice Recognition

Another data collection method that has gained considerable acceptance in the last few years is voice input or voice recognition technology. This allows the user to actually talk to the system and have the spoken words interpreted and recognized.

Most voice systems today are implemented through a plug-in interface card for a PC and sometimes an auxiliary unit connected to the interface card. A microphone is plugged into the system and spoken sounds (utterances) are digitized, compared to stored patterns, and, if matched, verified and passed to the application.

Pattern storage and matching use a lot of memory space and computer power, thus, the vocabulary must be limited in some way. The limitations can be either the number of utterances that can be recognized, the voice variances (number of different people), or, usually, both.

Voice systems can either recognize a very few utterances from most people or can have a much more extensive vocabulary (hundreds of utterances) but be limited to one person. Let's take the latter example first.

To set up a voice recognition system for Joe Schmoe, it is first necessary to "train" the system to recognize Joe's voice. The meaning of the utterances, which can be either words or phrases, are first entered into the system memory. In the training mode, the system will display these meanings one at a time and the user (Joe) will speak the word or phrase into the microphone. The system stores the pattern. Training usually involves several cycles of speaking the utterances, in a variety of ambient conditions (quiet room, noisy room, normal work environment) to allow the system to characterize normal variations of the pattern.

Once the training is complete, it's ready to go to work. These systems are remarkably adaptable for day-to-day variations that are typical of the human voice. Even the common cold and respiratory congestion are not often able to defeat the system's ability to correctly interpret the sounds. Most systems are capable of verifying the interpretation by either "speaking" back in a synthesized voice or displaying the interpretation on a screen within view of the operator.

The stored voice patterns can be kept on a cartridge, similar in appearance to a video game cartridge, or on a disk or diskette so that the recognition can be changed at shift-end or whenever is necessary.

Since the system doesn't really recognize words, but only matches patterns, accents or even foreign languages don't matter. If the interpretation is "1," it doesn't matter if the utterance is "one," "uno," "un," or "supercalifragilisticexpialadocious"—the interpretation is still "1."

The other type of recognition is not user-specific. This is a lot more difficult because of the large differences in pronunciation and voice sounds. The interpretation must be quite broad, therefore, the vocabulary must be very limited. A combination system could be used to take advantage of the strengths of both as in the following example.

A sales order system can be set up so that a salesperson can call in and reach the "universal" voice recognition system. If the system is trained for the numbers zero through nine, the salesperson can speak his/her ID number, one digit at a time and, when recognized, the system can then load that person's file with his/her specific vocabulary of several hundred words. The salesperson can then enter his/her order with verbal verification and prompting.

An easier approach to the same application would be to have the salesperson enter his/her ID number using the touch-tone pad on the telephone. This system is in use by a number of banks. A customer can call in and enter an account number and hear his/her balance from a synthesized voice.

Voice recognition is particularly useful when the person making the entry cannot easily free up one or both hands to pick up a pencil or bar-code scanner. This is also a very quick method for use in high-volume entries. Inspection tasks often include both restrictions. Voice is also appropriate when the person supplying the data is wearing gloves, has greasy or dirty hands, or is physically handicapped.

Battery powered memory units are also available. An interesting example of the use of voice is in the final inspection area of Ford's Taurus/Sable plant in Atlanta. Two inspectors walk past each car as it is about to leave the plant, and speak their observations into portable units. The data are then "dumped" into a PC controller periodically. The reports are specific, quickly entered, and are not delayed by writing and later key entry.

Early voice systems were very limited and required the speaker to clearly utter each sound or word individually. Today's systems aren't so limited, and can interpret "continuous" speech (sounds run together) much better than older systems could. The technology is improving, and more flexibility is certainly in the near future. Today's speaker-independent systems can recognize 10 to 20 words; speaker-dependent systems are capable of 600 words.

Voice, bar-code, and other data collection methods can be combined into a single system using the most appropriate collection method for each data station. The collection system controller (and the using application) receives the same kind of data stream, defined for the application, regardless of the originating device, although format conversion is sometimes required.

Advantages of Data Collection

The two biggest advantages of an automated data collection system are the ease of the reporting process for the users and the increase in timeliness and accuracy of the reporting. Obviously, anything that we can do to eliminate or simplify any activities that detract direct labor from the production tasks is a plus. Given that activity reporting is essential for getting the benefits from your planning and control system, the more timely and accurate the data, the better your system will be able to support you.

Installation of a data collection system should definitely impact the timeliness of transaction reporting. Since many systems today support on-line transaction entry, the business system database can be kept up to the minute for better status tracking and management visibility. Compared to the one to two day delay in updating that is typical with manual entry arrangements, this is a major improvement.

Accuracy improvements directly attributable to the collection method are limited to the elimination of those errors that are related to the handling of the data. Specifically, writing incorrect information, misreading the handwriting, number transpositions, lost transactions, miskeying, and the like are the types of errors that an automated approach can prevent. It is still possible for the worker to wand the wrong bar-code symbol, but the system might be able to detect such an error and reject the transaction on the spot.

It is a mistake to assume that the installation of a system will take care of all error sources. The workers must still be properly motivated to report accurately and not be punished for doing so (refer back to the first few pages of this chapter).

You may experience some improvement in attitude by virtue of the simplification of the reporting tasks and what is known as the "Hawthorn Effect." This phenomenon is named after a General Electric plant that experimented with working atmosphere by painting the walls in various areas of the plant different colors, to see if this would have an effect on worker productivity. What they found was that painting the walls, no matter what color, improved productivity. In just this way, the attention focused on the plant by the installation of a data collection system might improve performance outside of any direct impact that the system delivers.

Once again, if you have a fifty percent accurate reporting system, and automate it, what you end up with is a real-time on-line, automated, fifty percent accurate system. Attitude and motivation will determine your success.

It has been my experience that data collection equipment is difficult to justify on its own. When brought up before the finance committee for capital appropriation, the proposal will often fail. Companies seem to be much more willing to invest a half-million dollars in a new machining center than to spend fifty-thousand on data collection, for the simple reason that the system doesn't produce parts. The justification is indirect: more timely reporting, more accurate records, savings in direct labor time that is not currently measured (reporting is part of production, not separately reported). The only direct savings is key entry, and this is seldom enough to justify the investment.

Most of the systems that I have seen installed were purchased as a part of a new business system. As an incremental cost included with a much larger project, data collection can often ride the coat tails of MRP II or some other modernization program.

Data Acquisition

Many programmable controllers (PLC's, covered in Chapter 7) are connected to sensors which report process conditions for use in the control applications. If the controller is to keep the temperature at a certain level, it must have a temperature sensor to provide input as well as a control capability (output). Some controllers are capable of storing the process parameters (measurements) and/or passing them on to another system through a communications port. PLC's are also used to control Coordinate Measurement Machines (CMM's) which automatically take precise measurements of completed parts for QC purposes. Measurements can also be stored or communicated. These are examples of data acquisition. Also known as data logging, this definition can also refer to recording the measurements on a chart at the machine or controller. We will not discuss that type of data logging here. A commonly used acronym is SCADA for Supervisory Control And Data Acquisition.

Process monitoring, in addition to its use for direct control applications, is used for quality monitoring (covered in Chapter 9) and also for alarm generation. If a process must be kept within certain limits, and the controls fail to maintain the process within the acceptable range, then it is important to call the situation to the attention of the human operator or supervisor. Alarms can be directly connected flashing lights, horns, bells, or other mechanisms, or could be tied to a computer for display on a monitoring screen. A severe alarm condition might automatically shut down the process.

While data logging can, theoretically, be used for production activity reporting, there are some practical considerations that have prevented its use in this way, except in very rare instances. The data available from a controller could be production rate or count at a particular time, but the production monitoring system wants to know order/task identification, elapsed time, and count at some measurement point (completion of job, end of shift). Something would have to signal the data logger to take and send a measurement, and the identifying information must be added to complete the transaction. It is usually easier to manually report or use a terminal (bar-code or other). These considerations do not apply in continuous production environments where direct connection to plant management systems is more common.

If the logger is connected like a wedge-type reader device, that is, in line with the keyboard of a terminal, counts and rates can be inserted into a transaction by pressing a button of the controller to send the data. This method is a lot more practical than fully automated reporting. Measurement tools can also be equipped with digital encoders to operate as just discussed. The measurement is made and "inserted" into the transaction in lieu of key entry of the measurement number.

Plant Floor Systems

Computers on the plant floor include process monitoring and control systems (next two chapters), data collection devices, information distribution terminals, and some local activity control applications. There is a move to eliminate paper in the plant that has resulted in more widespread use of terminals and other display devices in the plant. There are products available now that will allow the electronic distribution of drawings and manufacturing instructions, as needed, to terminals in the work area. These terminals sometimes have limited ability to manipulate the graphic image, such as zoom and rotation.

Traditional MRP II systems will schedule and prioritize work to the work center/day level. If the work center contains a number of resources (machines), and many jobs are run each day, this may not be detailed enough for effective management of the resources and work flow. Distributed systems have been introduced that will accept a "download" of the MRP II schedule and allow management to the machine/minute level within the work center. While this level of control is not required by every manufacturer or every work center, it is an important capability in some situations.

When using this type of two-level control, the systems must be synchronized. Double entry of data is not an acceptable approach. Data entry would be required at the lower level (more detailed), and the distributed system should be able to summarize and "upload" to the MRP system.

With all of these devices sharing the plant floor area, communications networks become very important components of the systems. Typical organization would include separate networks for the application groups (control, order tracking, data distribution, data collection), but it is possible for PC-type devices to serve multiple functions, given enough memory, the proper interface cards, and broadband networks which can handle multiple signals.

Chapter 6 Review Questions

1. What is the accepted bar-code symbology for manufacturing? For retail package marking? For products sold to the government? For cartons of grocery items?

2. How many start and stop characters are there in each of the following symbologies? Code 39, UPC, Interleaved 2 of 5, Code 128.

3. Of the four codes discussed, which has the best (least) error rate, which has the worst (highest) error rate?

4. Why isn't OCR more popular? What's the problem with magnetic encoding?

5. Will the installation of a bar-code data collection system improve reporting accuracy?

6. How does an event differ from a transaction?

7. What are some reasons for using voice recognition?

8. Why is it hard to get funding for data collection?

9. What is data acquisition?

Automation of production machinery started with primitive controls that were added to weaving looms in the first years of the nineteenth century. Weaving was perhaps the most complex production process of the time, and the intricate patterns that were woven into cloth required precise changes of thread colors and often involved repeated patterns.

This earliest form of factory automation involved the encoding of loom instructions as a series of holes punched in wood or metal "cards." Mechanical fingers con-

7. Process Automation

trolled the movement of loom components based on whether they found a hole in the card, and therefore passed through to a more advanced position, or the absence of a hole which limited the movement of the finger to an earlier stopping point. These instruction sets — early programs — allowed the production of complex patterns in the cloth quickly, reliably, and repeatably.

The concept of a stored program or set of instructions dates back to Charles Babbage who first proposed an "analytical engine," a mechanical computing machine using punched cards as the storage medium for the instructions. Babbage was unable to build his machine because the manufacturing technology of the time was not able to produce the gears and cams to the required high level of precision. Work on this project did, however, help advance production capabilities and increase the precision of production processes significantly. Joseph Jacquard's use of the program cards in his loom in 1801 was the first practical application of the programming principle.

The simplest machine controllers in use today still use a series of punched holes, this time in a paper tape, to transfer instructions to the machine. The holes now confer electrically interpreted data rather than mechanical position as in the automated loom's "fingers." The paper tape-encoded information is passed through a reader where the holes in the paper (or lack thereof) control the making of an electrical connection (where the hole is) or absence of a connection where there is no hole. The connection/nonconnection provides binary information (ones and zeros) to a controller which translates the signals into machine instructions such as "move in 3 inches," "lower the cutting head one-half inch," "raise the cutting head," "move right one inch," etc.

The general name for this kind of basic machine programming capability is Numerical Control (NC). The example used here is a one-way communication: from the controller to the mechanism, and the machine simply follows the instructions one after the other. This type of control function is called sequential control. The operator is responsible for placing the work in the proper position, usually using a fixture of some sort for accurate positioning. The operator also aligns the tool and sets the machine at the specified starting position. The operator then initiates the program and the tool follows the instructions. Once the sequence is completed, the operator removes the work piece and probably inspects it to be sure everything went according to plan.

A simpler method, that is sometimes used in a lathe, is a purely mechanical approach. With a tracer lathe (sometimes called a follower lathe), a pattern is clamped into a holding jig and a mechanism traces the pattern, mechanically or pneumatically transmitting the movements to the cutter. A copy of the pattern is thus produced in the work piece.

Material processing activities can be categorized into two general areas: material forming and material removal. As the names imply, the shape of the work piece can be changed through a forming operation such as a forge, press, or brake, or some of the material can be removed by a process such as drilling or sawing.

Sequential control capabilities, such as those introduced above, are most frequently applied to mills, lathes, and drilling machines — all material removal processes. In a mill, the cutting tool is usually held stationary in a rotating chuck while the work surface is moved left and right (X-axis), in and out (Y-axis), and up and down (Z-axis). Mills are used to shape a piece, cut grooves and notches, and/or smooth a surface by removing unwanted materials from the work piece. The foregoing description is for a vertical mill, so-called because of the orientation of the axis of the cutting bit. Horizontal mills, with the cutting tool axis mounted horizontally, are another often-used style.

In a lathe, the work piece is rotated (spins) and the cutting tool is moved across the piece (X-axis) and in or out (towards and away from the center of rotation, Z-axis) to produce rounded shapes. Lathe cutting tools can be set up for cutting the outside surface or cutting into the end of the piece (inside diameter). In a drilling machine, the drill bit is positioned over the proper position by either moving the drill or (more often) the work piece. Motion over the surface is in two axes (X and Y); and drill vertical travel, the Z-axis, is also controllable (hole depth and cutting speed).

Programming a mill, lathe, or a drill involves specifying the movements of the piece and/or tool, in the proper sequence, the proper direction, and the proper amount, to perform the cutting action desired. This is known as sequential control. While automation has been applied to material forming processes, it is most often in the form of associated systems for handling the materials — placing and removing the work piece. This type of automation is addressed briefly later in this section, and more fully in the sections on material handling and work cells in Chapter 8.

The beginning of sequential control of machines can be traced back to "clockwork" mechanisms — arrangements of gears and cams to perform a sequence of activities. These completely mechanical systems were designed to perform a single task or set of tasks, and could not be reprogrammed to do anything else without rebuilding the mechanism. Mechanical controls were replaced in the early part of the twentieth century by groups of relays controlling electricity to motors, solenoids, and electrically actuated valves. Programming for the relays was "hard-wired," meaning that the electrical connections were set up for a certain task. The wiring, and therefore the programming, could be changed, however, a lot more easily than rebuilding the machine as was required with mechanical arrangements.

Sequential relay controls can be actuated by a timing mechanism for "open" controls which will repeat the sequence, the same way, each time. Open controls are those that do not involve feedback. A good example of this kind of control is the household washing machine. The same sequence is followed whether or not there are clothes in the tub, even if you forgot to put the detergent in, even if the clothes come out as dirty as they went in. If the operator checks the process and adjusts it, you now have a "closed" control loop with feedback. If the washer is checked mid-cycle and the operator then turns back the timer to add some wash time, a control loop cycle has been completed: measure, evaluate, adjust if necessary.

The washer feedback loop required human intervention since the system did not allow for feedback. Obviously, some systems have the feedback mechanism built in. In modern systems which use electrical or electronic control, the feedback process makes use of a transducer of some sort. A transducer is an electromechanical device that produces an electrical signal which is relatable to a physical reality. A transducer can be a simple switch: when the refrigerator door is closed, the switch contacts are opened which turns off the light. A temperature sensor can complete a circuit when the freezer compartment rises above 30 degrees, thus turning on the chiller. Alternatively, a sensor can produce a voltage or current proportional to the temperature, and the control system can interpret the signal and decide when to turn on the chiller mechanism. Transducers come in all shapes, sizes, and capabilities.

On the other end of the loop, the opposite is required: the ability to change an electrical signal in a physical reality. This takes an actuator. A relay uses one electrical signal to control another (so does a transistor, although in a very different way). Other commonly used actuators are solenoids which use electrical current to push or pull a magnetic plunger, motors which turn current into spinning motion ("stepper" motors are motors in which the motion can be controlled in precise increments), and electrically activated pumps and valves which are special-purpose motors and solenoids, respectively.

A typical (simple) relay programmed task using feedback might involve the control of the liquid level in a tank. The system might consist of two sensors (to indicate when the tank was full and when it was empty), the relay controller, and a pump. When the "empty" sensor indicates that the tank is indeed empty, the signal would be used to start the pump operation. When the tank is filled, the "full" sensor signal should cause the pumping action to stop (Fig. 7-1).

The logic could be simply stated: if the tank is empty, turn on the pump; if the tank is full, turn off the pump. If neither indication is present, don't do anything. This logic can be illustrated on a chart called a "ladder diagram" which describes the relationships between switch positions (sensor states) and the resultant activities (system output, usually voltage or no voltage).

Obviously, Fig. 7-2 is an extremely simple example. The symbols on the left indicate the sensor state active (on), and the symbol on the right means that the output should be energized or deenergized accordingly. Other symbols indicate that the sensor is off or deenergized --|\|--

A more complex representation might include a power switch, other sensors, interdependent conditions, etc. Four basic symbols are all that is needed in ladder diagramming since the process is essentially binary: a signal is either on or off (energized or deenergized), and the output signal is either to energize the actuator or deenergize it. The symbols are:

--| |--   input on (energized)

--|\|--   input off (deenergized)

--( )--   energize output

--(\)--   deenergize output

The ladder diagram became a standardized way of illustrating relay logic because it was easily relatable to the actual wiring of the relays. The makers of the first programmable controllers in the United States used ladder diagramming as a way to describe the computer logic used in the controllers. Because of the familiarity of the tech-

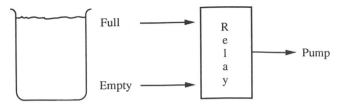

Fig. 7-1. Tank fill control.

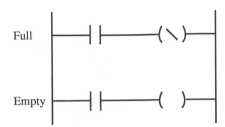

Fig. 7-2. Ladder logic diagram.

nique, controllers were readily accepted here as replacements for relay boards. European controller manufacturers did not adopt ladder programming, and acceptance there lagged considerably behind the U.S. There is a valuable lesson here for anyone introducing new systems or techniques to the plant floor, or anywhere else for that matter.

While ladder logic is not required for programming modern controllers, many of the programming languages available today support a form of ladder logic entry method to allow users brought up with this way of thinking to easily relate to the programming process. Ladder diagrams work very well for closed (feedback) control systems in which an output action is dependent on input conditions, but they are very cumbersome when used to describe sequential control applications.

Control systems can be special-purpose or dedicated to support a particular task or type of activity, or they can be general-purpose to allow more options in the programming of the controller to different tasks and applications. Dedicated industrial controllers are called process controllers (the abbreviation commonly used is PC), and are often supplied with machines from the original manufacturer. I will not use that abbreviation in this book since PC is also used for personal computers, which are finding their way onto the shop floor in increasing numbers today. Programmable Logic Controllers (PLC's) are the general-purpose controllers which tend to be more flexible and handle more tasks than dedicated controllers. PLC's sometimes are supplied with machines from the factory, or they can also be purchased separately for installation at the plant on existing machines (equipped with the necessary actuators and/or transducers) or for controlling multiple machines which must perform functions in coordination.

Today's programmable controllers are becoming more like general-purpose computers than the traditional, dedicated, wired-in machine controls of the 1950's. Personal computers can also be used to supervise activities in much the same way that PLC's are used. In fact, in this area, just as in general computing, the lines which traditionally separated different machines and capabilities have all but disappeared. Dedicated controllers are becoming more flexible and capable, PLC's are performing traditional computer tasks, PC's have moved onto the shop floor, and networks tie all of them together in integrated systems.

Dedicated controllers are designed to perform specific tasks associated with a particular machine or machine type. The tasks can include both sequential operations and feedback systems for process control. On a typical machine tool control, there must be a capability to provide instructions (the program) to the controller which is most often a sequence-type program. The incorporation of feedback allows position sensors to report back to the controller when a sequenced movement is completed, and to provide a measurement of the physical position of the machine's actuators to confirm that the requested motion was completed successfully.

Once the controller is installed on a factory machine, the operator must supply the programs. For a simple drilling application, it would be necessary to define the position of each hole (X and Y coordinates), the hole depth, and drill motor speed. Most controllers provide the capability to enter or modify the program right at the machine, and include either a simple read-out display or a video screen. If feedback is included, the display will show the current program step, the tool position, and operator messages such as "operation complete," "load material," or warning or error messages.

With the advent of better (faster, more capable) controllers, while the basic tasks didn't change, the number of steps, and the complexity of the movements, could be expanded. More flexible machines can now be controlled allowing more choices and greater capabilities. An example is the ability to change tools on the machine.

With a simple lathe, a tool and work piece are mounted and aligned, then the program directs the motor speed and tool movement. If the contours of the piece require several different cutting tools, the operator must stop the process at the appropriate time and switch tools. The simplest approach to solving this problem is to use a turret to hold four or five tools (some turrets hold up to eight tools). The controller now has an additional task: to turn the turret to switch tools. An expansion on this idea is the carousel which can hold up to fifty tools or more on some machines. Additional mechanisms on the machine allow the carousel to present the proper tool to a manipulator which removes the tool from its position on the carousel and places it in the business end of the lathe (or other machine). When it's time for a change, the old tool is replaced and the next tool moved into position. Tool changer mechanisms often have two tool holders so that the next tool can be prepositioned for quick changeover with the previous tool replaced in the carousel after the new one is put in position. Tool change in modern machines can usually be accomplished in several seconds.

Other advancements in machine technology include the incorporation of more functions into one machine. In a lathe, a second tool head can be installed to provide inside diameter cutting capabilities or a second outside diameter cut at the same time or more quickly in sequence with the first cutter. Another motor installed in one of the cutting heads reverses the original orientation and allows the turned piece to be grooved, milled, or drilled while still in the lathe (work piece not spinning). In mills, more freedom of movement as well as tool carousels, additional cutting heads, etc., can provide more capabilities. For example, a rotational axis can move the piece in different ways, and a two-position work holder (pallet) allows the operator to remove a finished piece and load a new one while the machine is working. When the work is completed and the new piece is loaded, the work holder is "swapped" end-to-end and the process restarts.

An example of a (dedicated) combination sequence-control application would be the control of the freeze-drying process. A freeze-drying cycle requires that the temperature of the drying chamber be lowered to a certain level for a certain length of time, then lowered to a different temperature for a period, then brought back to ambient (room) temperature. This is obviously a sequence of activities, but it is also necessary to read and control not only the temperature, but also vacuum (pressure).

A microprocessor based controller can be built into the drying apparatus or added on to perform the sequence and also control the temperature and vacuum pump. The controller may be supplied with a preprogrammed drying cycle, and the operator need only enter the critical parameters such as the temperature to be maintained at each stage and how long each stage must last.

Controllers can be programmed by directly entering the instructions at the machine or by loading from a tape. The more advanced controllers often have the ability to store multiple programs which can be recalled as needed. Because these controllers contain a micro-processor with associated memory and storage capabilities, they are called Computerized Numerical Controls (CNC).

When the programs are stored on a separate computer — often a general-purpose computer such as a personal computer (PC) or a workstation — and "passed" to the machine controller when needed, rather than stored at the machine, this is called DNC, which stands for either "Direct Numeric Control" or "Distributed Numeric Con-trol." The computer that stores the programs is often connected to more than one machine controller and acts as a repository for many programs and several different machine types.

Another general class of automated machines performs assembly operations. All previous discussions involved reshaping a single work piece. Assembly machines, as the name implies, combine separate pieces together into a new entity which could be a part or a finished item.

Automated assembly machines have been around for many years, but only recently have they become easily changeable from one prod-uct to another. Dedicated (single-purpose) assembly machines tend to be large and expensive and can only be justified for a high-volume activity. The capital cost and the set-up or changeover difficulty dictate that the machine must run for a significant time on the same product to justify the fixed costs involved.

A commonly seen type of assembly machine is a packaging op-eration. Typical in consumer goods products, packaging machines can often perform the same operation many thousands of times a day. To change from one product to another often requires shutting down the machine for several hours while changing products, pack-ages, labels, jigs, and fixtures.

Since assembly-type machines often involve a large number of coordinated movements, programmable controllers are often used to manage the activities of the various parts of the operation. Controllers can also record production rates and quantities, and even measure various parameters such as package weight. With a controller-man-aged machine, changeover of the instructions is quite simple: either call up a new program from memory or down-load it from a computer or, worst case, enter the new program from the operator station. Mechanical changeover of the machine itself is usually a far more critical consideration.

The newest development in this area is the use of sophisticated material handling devices such as robots to perform assembly activ-ities and move parts and assemblies from one work station to the

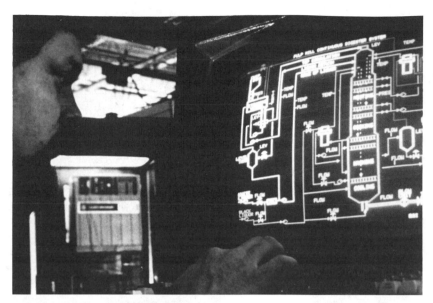

Fig. 7-3. Color graphic displays assist the operator of sophisticated monitoring and control systems. (Courtesy Allen-Bradley Co., Inc.)

next. The advent of these new, more flexible machines allows quicker changeover from one product to another because more elements of the activity are programmed rather than being based on mechanical components. It is a lot quicker to call up a new program than to disassemble a fixture and change it to another. While mechanical components are still part of any assembly operation, more flexibility and faster changeover can result from more programmable elements.

The ability to change products quickly is increasingly important to a manufacturing company in today's markets. Historically, "economy of scale" favored the high-volume producer. As discussed, mechanical automation is expensive and hard to justify for small production quantities. Therefore, only those products and producers that could support high volume could afford to buy and operate automated machinery which could reduce the unit cost to produce. Because of this relationship between high volume and lowered costs, flexibility is lost. If demand patterns change, the high-volume manufacturer is now at a disadvantage because his changeover costs are so high.

The trend in recent years has been toward more specialty products, that is, more variety at lower volumes per product. Everyone seems to be searching for market "niches" rather than a product with more universal appeal. Unique products (without a lot of competition) often produce higher margins. To be successful at lower volume, fixed costs must be minimized. Automated machinery can lower production costs, but only if the set-up costs are not prohibitive.

Programming

Programs can be developed right on the machine or "off-line" on a computer. For programming at the machine, the operator might use a keyboard or other entry device to enter the instructions one after another, or, on some machines, a recording technique could be used. This is a simple method of programming, and one that is used on older machines and on machines with complex positioning capabilities such as robots (covered in the next chapter). In the recording method, the machine is positioned as it would be while performing the job, and each key position or motion change point is recorded using a type of "enter" button on a console or remote control. The recording technique has limitations, but can be a useful capability in some applications. See Fig. 7-4.

Developing programs on a separate computer requires the use of a software program or language to interpret the instructions entered by the programmer and convert them to commands that the machine controller can understand and execute. There are many such programs available, each with its own characteristics and capabilities. Some allow the use of ladder diagrams to describe the process, converting these diagrams to programs which can be used to drive the machine through its controller. While there is no "standard" for machine programming, the programming language called "C" seems to be becoming the dominant language for this type of work on PC's and workstations. Dedicated NC programming computers generally use proprietary languages and programs. A general-purpose NC program development system will create the machine instructions, but a supplemental program, called a postprocessor, is used to translate the NC program into one that can be successfully executed on a particular machine or controller type. Postprocessors are available for most widely used controllers.

It is also possible for an engineering computer system to develop machine instructions from the graphical description of the part that is developed in the Computer Aided Design (CAD) system. In larger, more powerful CAD systems, additional programs are available that can read the mathematical description of the surfaces of the part and generate a complete set of machine instructions (tool-path) that can be passed to the controller. This topic was discussed more fully in Chapter 4.

Fig. 7-4. Material handling robot being programmed with a hand-held teach
pendant. (Courtesy Cincinnati Milacron Industrial Robot Division.)

When information is to be passed from one system to another,
there must be a definition of what that information will look like so
that the information will be usable when it arrives. Unfortunately,
there is no universal standard for data communication between engineering and the plant, or even within the plant itself. There are
many vendors, and until recently there was very little interest in
agreeing on standards. The dominant attitude seemed to be that a
unique protocol would encourage users of the vendor's products to
buy all associated products from the same vendor.

This strategy worked for a time, but a new market developed for
conversion products and interface equipment to allow the customers
to select equipment based on capabilities, not just compatibility. As
things developed, it became clear that the customer would find a
way to link various vendors' equipment together, and that the companies that supported this goal would be well received. Suddenly,
everyone is now an integrator. "Open" systems committees are being
formed to set some ground rules that, while not "standards," at least
establish guidelines and parameters that make interconnectivity much
easier than before. Also, many companies now offer "enablers,"

which are software tools that can be used to develop interconnections (not physical wiring connections but data communications interconnections) between equipment using different protocols and languages.

In the plant floor area, communications facilities must be established between controllers that work together, and among controllers and computers that work with controllers, to either provide direction or to hold and transfer programs. Direct connection works well when only a few devices are involved. Interconnection between devices requires that each device be equipped with a communications port, usually a serial port. Serial refers to the sequential nature of the data transfer, one bit at a time, in a series of bits which makes up the message. The alternative to serial is parallel communications, which is faster because more bits of information are transmitted at a time; but parallel requires more wires be run between devices.

Serial communications is usually accomplished using standard RS-232C or RS-422 interfaces. These standards define the connections, signal characteristics, etc. Laboratory instruments and measurement devices sometimes use a different interconnection standard known as IEEE-488 or General Purpose Interface Bus (GPIB). For more devices, more flexibility, and higher speed and capacity, many companies are now using Local Area Networks (LAN's) for device communications. Common LAN's in use in the factory environment are Ethernet, developed by Xerox and Digital Equipment Corporation, and Manufacturing Automation Protocol (MAP). Another networking arrangement which is becoming increasingly common is IBM's Token Ring network. LAN's also provide for connection to nonfactory systems such as corporate data processing systems. LAN's and protocols were discussed in more detail in Chapter 4.

There are many network systems available on the market. The first two of the three just mentioned are not commonly found outside the factory. PC networks in the office use IBM's Token Ring, Novell's Network, and others. This illustrates the basic problem in integration: products from various vendors were developed for specific markets and are incompatible with each other. To link an office LAN with a factory LAN can be a difficult undertaking. Linkages between compatible LAN's is accomplished using a "bridge." If the LAN's use different protocols or standards, then a "gateway" is required to translate the signals as they pass between the networks.

The LAN defines the physical connections (wire type, voltage and current levels, data transfer method and speed). The LAN vendor may or may not specify the protocols to use. Protocols are standardized formats, data encoding techniques, communications procedures (handshakes, identification codes), etc. Each LAN has its own standards — both physical and data. Communications systems and what is contained in a protocol were discussed more fully in Chapter 4.

LAN's fall into two general categories: baseband and broadband. Baseband LAN's move the signals around the wires using audio frequency modulation and the signal is relatively low frequency and low speed. Broadband LAN's use modems to impose the signal onto a radio-frequency carrier, which carries it through the network at much higher data rates.

Baseband networks can use twisted-pair cabling which is relatively inexpensive but limited to the single task of carrying the network data. Broadband LAN's require a special kind of coaxial cable which, although more expensive, can also carry multiple signals including video throughout the plant. There are variations including baseband on coaxial cable such as is used by Ethernet.

Networking has become a very big business in the last few years, and the selection of an appropriate network must consider the number of devices, their physical locations, the type of signals to be handled, future growth plans, and the compatibility of existing equipment. There are network experts available who should be consulted not only on the selection but also on the design of the network.

Once the network is selected and installed, you still have the challenge of making the devices (computers and controllers) understand each other. For plant-floor applications, the determining factor has to be the machine that will be running the programs. Whatever method is used to generate the programs, no matter what the storage and transmittal considerations, when the program arrives at the machine controller, the machine must be able to accept it as received. If there is some incompatibility along the way, it must be solved before sending the instructions to the machine controller or some place en route.

Conversion programs are generally available today from and to most commonly used formats. Some conversion routines are available from vendors of the program generation products. There are also independent companies that sell conversion utility programs.

The latest direction in the development of plant interconnection and integration is the move toward CIM standards and products to 1) encourage manufacturers towards compatibility through "open" or public standards, 2) provide development tools to simplify the translation and transfer of information among otherwise incompatible systems, and 3) provide centralized functions to hold and transfer information between participants in the CIM system. One significant effort in this direction was the announcement by IBM in October of

1989 of a "CIM Architecture" and related products in an attempt to establish a de facto standard for the industry.

Other Considerations

Once the capability to perform the manufacturing function has been established, it is necessary to bring this capability together with the need for the product and the availability of the materials to perform the job. Developing the production schedule and assuring availability of materials is the responsibility of the business and planning system (Chapter 5, Manufacturing Resource Planning). It is also necessary to provide manufacturing instructions and priority information to the line managers; once again, the development of this information is within the realm of the planning system.

In return, the production area must provide activity reports to the planning system for progress monitoring, status tracking, efficiency and utilization measurement, and actual job costing. In most companies, this two-way communication between the planning system and the plant floor is either completely manual (paper shop documentation and work lists, manual activity reporting) or at least includes a human in the loop. Some automation is available in support of activity reporting in the form of bar-code reporting and other systems, as discussed in Chapter 6. In the distribution of instructions and priorities, some systems have been announced to present this information as well as drawings and engineering data on plant-floor terminals or display devices. This area is still in its infancy at this time, but it should develop rapidly as in-plant networks become more common and as technology provides us with better display devices and improved system and software compatibility.

It used to be an accepted fact of manufacturing life that the way to reduce unit cost was to increase volume. Because there are both fixed costs and variable costs associated with any manufacturing activity, it is logical that a lower unit cost will result from spreading the fixed costs over a larger production quantity. Since most "overhead" costs (things like heat and lights and building depreciation) can be assigned to a cost per hour or day of operation, the most visible fixed cost is machine set-up time. With a given set-up cost, the larger the production run, the smaller the amount each part has to absorb.

In most equations there are at least two factors; in this one, there are set-up cost and run size. The obvious alternative to longer runs is shorter set-ups to reduce total cost. Automation of production machines has had a dramatic impact on set-up. With a large or complex machine, it is not unusual to expect a changeover to take hours or sometimes days. During this nonproduction time, the meter is still running on the cost per hour for the machine. Heat, lights, and depreciation still apply even when no parts are being produced.

Obviously, then, any reduction in set-up time should reduce unit cost to produce for a given run quantity, or should lower the economic run quantity to yield the desired unit cost.

A manually operated machine might have a short set-up time. A change of tools and/or jigs, clamps, holders, etc., might be the sum total of the change. A programmable machine must have a new program loaded along with the tool change. Basic NC machines would require that the program be loaded each time, either from an entry station or from a tape. More advanced machines might hold multiple programs in memory which can be "called up" as needed. This would be faster than loading each time. Finally, DNC allows a centralized computer to hold the program and send it to the machine controller as needed.

No matter the loading method, when a program is first developed, it must be verified (tested) on the machine with the tools and (usually) work piece in place. Even off-line developed programs sometimes need minor corrections at the machine. When a tested program is reused, verification is sometimes appropriate to compensate for slight differences in holder or tool position.

In any case, some changeover activity is required, and the quicker this can be done, the better. Machines with a high level of automation can, through the use of tool changers and flexible work holders, often reduce the changeover time to minutes. There is a common acronym for quick changeover: SMED, Single Minute Exchange of Dies, which is a goal in many automated factories. There is nothing "magic" about one minute, it is simply an aggressive target for minimum changeover time.

The biggest impact, today, of minimum set-up requirements is flexibility. As world markets have developed and become more competitive, it is becoming less important to develop high-volume capabilities and economies of scale and more important to be able to target products at niche markets and be able to produce products efficiently in short runs so that product mix can be managed more closely. Toyota and other Japanese companies have gotten tremendous publicity for their manufacturing triumphs using many techniques—one of which is this focus on flexibility. Instead of making four-door Corollas one day, two-door models the next, and wagons next week, production schedules reflect a mix of products each day. "Every product, every day" is a kind of watchword in flexible manufacturing plants, and this allows production of only those products that are needed rather than those that are convenient to make.

In Lee Iacocca's first book, he talked about Chrysler producing large volumes of cars and letting them pile up in the finished goods lot until they had to do something to move them. They got into the habit of offering "deals" to the distribution channel so often that the dealers would wait for these "fire sales" because they knew they would inevitably come. This operating policy did not do much for Chrysler's profitability. Most industries cannot afford such luxuries anymore (if they ever could). Markets are much more diverse and particular than they used to be, and the plant that can fine tune production and change products quickly has a big advantage over its competitors.

All of this discussion of changeover assumes that there is some sort of plan. Someone must decide what is to be made and when. Once the production schedule has been developed, the instructions are sent to the plant floor for implementation. This information is usually passed to the operator on paper. Often, the day's schedule and priorities are written out and distributed at the beginning of the day. The foreman or work center supervisor decides on the actual sequence of activities (which jobs to run next) based on the schedule and his/her own judgement, considering factors such as machine capability and availability, management priority, and personal preference.

The biggest risk for the integrated system at this stage is the motivations of the person doing the sequencing. No matter what the system has determined for priorities, it is the judgement of the people on the plant floor that really determines what gets done. If the motivation system for these decision makers is not tied into the system-generated priorities, the result will be affected. There is no substitute for judgement and experience, and there must be room for these in your operational system; but it is also necessary to properly motivate them. A common error in implementing integrated systems is to overlook this consideration.

Often, prior to automating, the sequencing of work in the plant or at a particular work center will be influenced by either an incentive pay program or some other reward/punishment system tied to some general company goals. Typical goals are the number of units shipped or dollar value shipped per period (usually a month). If this orientation is not changed, it will work against the integrated system.

I can remember, as a child, seeing my father come home from the plant on a Friday with "work slips" in his shirt pocket. When I asked about them, he told me that they were production counts (labor tickets) for that day's work that he was going to turn in on Monday. His incentive was on a weekly count of items produced, and if he had had a good week but saw some difficult work in the queue for next week, he would hold out the tickets to "smooth out" the production reports and therefore his paychecks.

This is a rather dramatic example, perhaps, but illustrates how people will respond to the way they are being judged. In your plant, it may be much more subtle but, let me assure you, if the incentive is not tied directly to the priorities of the overall production plan, you have a problem.

A coordinated production scheduling system (part of MRP II) will develop priorities for each production process and/or step that reflect the imperatives of the master schedule (end product ship schedules or inventory goals). It is important to find a way to motivate production personnel to adhere to these priorities. At the same time, however, there must be enough flexibility in the system to allow these people to exercise their judgement.

When a supervisor is presented with a prioritized job list for the day, he/she should be free to make adjustments to make the best use of his/her resources. If he/she can save nonproductive (set-up) time by delaying a high-priority job for a short time, or advancing a lower priority job, then he/she should have the freedom to do so. What makes this work is his/her tie-in to the master schedule priorities. As long as the schedule priority remains the top consideration, other factors can be manipulated to produce the best overall result.

A factor that has been recognized as less important in many situations than was previously thought is machine utilization. Before we became enlightened, we strove to keep utilization as high as possible — a busy machine is a happy machine. In the world of Just-In-Time, we have come to realize that it is better to have a machine sitting idle than to have it producing unneeded parts.

As in most other situations, things are not as simple as they might at first appear. Absorbing fixed and overhead costs over as much product as possible is still a sensible goal, but minimizing inventory is also important. This is especially true as product life cycles are becoming shorter and shorter, and the extra parts you produce today to keep the machine busy could end up on the scrap heap tomorrow when the product that used them is discontinued. Once again, management judgement must be applied to find the best compromise between producing only what is needed now versus keeping resources busy.

Plant floor access to management systems and information can play a role in making these critical decisions. Wider distribution of management information, especially production plans and priorities, allows decisions to be made at a more detailed level, allowing more flexibility and closer management of individual resources.

Of course, information is only useful insofar as it is timely and accurate. This requires good reporting systems from the plant floor to the management system. Thus, the supplier of the (plant/job/resource) status data to the management system is also the chief beneficiary of the resulting priority and scheduling information. This is the essence of CIM and MRP II: systems that tie together disparate portions of the organization through timely distribution of information that can enhance decisions and management control.

A key to success in distributed systems is the full involvement of the users in the entire process, i.e., taking an "ownership" attitude toward their portion of the system. When new systems are implemented, there are changes in the way business is conducted. Decisions are now based on the new system and the decision maker must be confident that the information is up to date and accurate. The only way that the required level of confidence can be achieved is to have the users of the information completely immersed in the collection and reporting of the data. They must take responsibility for the portion that directly affects their function — from start to finish.

Accuracy of reporting is crucial to the effective operation of the system. If the people doing the reporting are not properly motivated, the resulting errors will destroy the integrity of the resulting management information. For example, let's say that a production worker reports 95 pieces completed and 5 scrapped on a production run of 100 pieces. If the supervisor beats on the worker because of the scrap, the worker will find a way to not report it. It is amazing how readily people can find a way "around" any kind of controls if it makes life less painful for them. The worker in this example is highly motivated to hide the scrap information, and will find a way.

The opposite situation also holds true. If there are components left over, they should be returned and reported. If this is made difficult or painful, it won't happen. If component scrap (or resupply) is a bother, people will build up a private stock of extra parts to use when a shortage or scrap occurs. What do you suppose that does to any efforts aimed at inventory accuracy or inventory reduction?

The order being reported is obviously going to be short of the planned quantity or have a cost variance. O.K., this is not desirable, but at least we now know as soon as the situation occurs, and we can take corrective action as soon as possible. Late reporting, or only finding out at a later operation or the end of the job, only reduces the reaction time available. The impact of not reporting accurately goes far beyond the order that is being worked, however. Activity reporting is the feedback loop that allows us to validate bills of materials and routings and to keep inventory records accurate. If there is information withheld or inaccurately reported, we have lost the ability to adjust our expectations in order to plan much more effectively the next time we need to make the same part or a similar part.

The business and planning system doesn't really care about the causes of scrap or measure quality parameters; it is only concerned that the quantities required are available at the proper time. Simply reporting that losses have occurred is enough to support the cost collection, inventory control, and planning functions. Quality monitoring can also make use of this quantity information in developing statistics to be used in the planning system, but quality control is usually concerned with much more than just the numbers.

All manufactured products must comply to certain expectations in regard to size, shape, color, configuration, etc. Measuring these characteristics or parameters and reporting the results to a statistical analysis function is Statistical Quality Control (SQC). When the measurements occur during the processing and are used to make adjustments to the process to keep it within the desired limits, it is called Statistical Process Control (SPC). Both of these topics are addressed in a later chapter. Parametric measurements are frequently an additional reporting task imposed on production personnel. In some situations, quality inspection is a separate, defined process step. The finished or partially finished parts are routed to a separate inspection function, and are either accepted, sent back for rework, or rejected. More often, today, direct production workers are becoming more involved in quality measurement and process control. Measurements are made at the production step, and problems are detected as they occur.

Some automated machines include measurement and reporting capability whose data can be transmitted to a central control facility, or may display the results directly on plant-floor devices that warn of out-of-control conditions. Other machines are actually self-correcting based on the measurements.

Chapter 7 Review Questions

1. What is closed-loop numerical control?

2. Why is ladder logic still in use today? Is that a problem?

3. What is the difference between a PC and a PLC?

4. How can a PLC be programmed?

5. What is DNC?

6. What information is provided to the plant floor from engineering? From the business and planning area?

7. What information is provided *from* the plant floor?

8. How can unit production cost be reduced?

9. What is the biggest impediment to implementation of automated control and management systems?

10. Is it always good policy to keep all machines busy as much as possible? Why or why not?

The next step after automating a process itself is to extend the automation to the loading and unloading of the machine so that it can be operated unattended or with minimal attention. To do this, the materials (work pieces) must be moved to the location of the machine, placed in the work position when the machine is ready, and, when finished, removed or moved to the next operation. On the plant floor, these material handling systems consist primarily of conveyors of various sorts as well as robotic devices. Getting

8. Work Cells, Robotics, Material Handling

the parts from the stock areas to the shop floor involves Automated Storage and Retrieval Systems (AS/RS) and Automatic Guided Vehicles (AGV's). With the exception of simple conveyors, all of these systems use programmable controllers to hold and execute the programmed movements.

Robots

Despite the science-fiction image of the robot as a human-like mechanical creature on two legs, the thing that distinguishes an industrial robot from other types of automated machinery is the ability of a robot to move an object through space. A robot is essentially an arm-like device with the ability to pick up an object, move it, and place it in another position or orientation. The object moved might also be a tool such as a welder or paint sprayer. What makes a robot a robot is the arm-like nature and freedom of movement.

A robot's ability to move often includes more axes than a fabrication or assembly machine. As shown in Fig. 8-1, this robot can move whatever is attached to it to any position within a fairly large area. The area within which the robot can move its manipulator arm is called the envelope of operation, and different robot types have different size and shape envelopes.

Because of the number of axes and the potential complexities of the movements, a robot is much harder to program than machines with just X, Y, and Z axes. There is also the problem of precision of movements and bending of the members due to stresses imposed by different weights suspended at the end of the arm. Because of

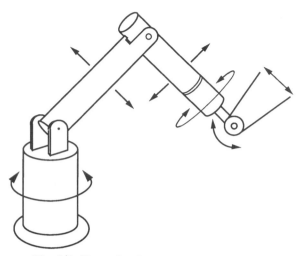

Fig. 8-1. Example of robot axes of movement.

these factors, programming robots is often done using a recording technique. The arm is positioned using manual controls (typically an attached "box" called a teaching pendant), and each successive position is recorded by the controller. The computer develops the instructions to move the arm from place to place.

Not all robots resemble the above illustration. Many different mounting options and mechanical arrangements are possible, and are used in different applications based on the size and weight of the objects to be moved, the range of movement, precision of placement required, and other factors. Movement can be driven by hydraulics, pneumatics, or electrical motors. Electrical motor drives provide the highest precision but cannot generally handle as much of a load as hydraulics or pneumatics.

The simplest robotic devices, called "pick-and-place" robots, incorporate no feedback mechanisms. These devices are used to move objects whose starting position is precisely known. If the object is not where it should be, the robot will perform the movements in exactly the same manner as if it was there. Obviously, this restriction limits the applicability of pick-and-place robots, however, this type of device is widely used because of its relatively low cost. Nonfeedback robots (open loop) are also subject to positioning errors caused by mechanical effects such as backlash and imprecise motor control.

A slightly "smarter" robot will use servo-motors for more precise movement. A servo-motor includes a feedback mechanism, such as a shaft encoder, which is a device to "read" the actual position of the motor's shaft usually using an electro-optical device. Servo-motors are self-correcting in that the motor can be used to move the joint or member to a measured position. The encoder signal is used to determine when to shut off the motor (when the desired position is reached) rather than running the motor for a predetermined amount of time.

Robots can be used to place a work piece in proper position for machining or other operation. This application, of course, often also requires that the piece be picked up after the operation is complete and placed elsewhere. The raw part must also be brought to the correct staging position (so that the robot can pick it up), and the finished parts removed from the area after the robot has put it down. These can be accomplished using moving belts or trays or other mechanisms.

In assembly applications, the arm will typically pick up a component, position it, and insert or place it on the assembly. Often the assembly will move into position on a conveyor, the robot does the placement, then the next assembly moves into position. Robots are much easier to reset for a different task than mechanical assembly machines. While the purchase cost might be substantial, the ability to change over easily allows smaller economic run lengths and therefore more flexibility in the manufacturing process.

A robot can also hold a tool such as a welding head or paint sprayer. Once again, the range of movement and programmability are key to these applications. See Fig. 8-2.

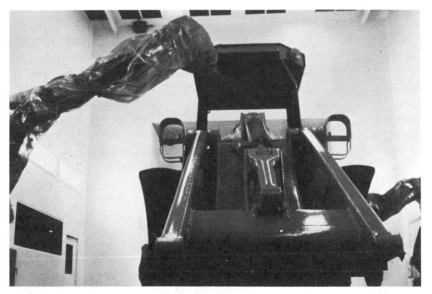

Fig. 8-2. Special-purpose robots can apply an even layer of paint, even in hard-to-reach places. (Courtesy Allen-Bradley Co., Inc.)

Current robot development, sometimes called the second generation, is focused on more sophisticated feedback systems such as pressure sensing and vision. If a robot can "feel" the object it is holding, and only exert enough pressure to pick the item up (and therefore not crush it), delicate items can be handled. Sensors in the "fingers" can provide feedback to the drive mechanism to stop movement at a certain pressure. Additional control system commands are used to specify the pressure to be used in each application.

Vision systems can help direct the movement of the device and reduce the need for precise placement of the parts that are being assembled or fabricated. As an example, a robotic seam (arc) welder without vision must follow a programmed path exactly. If the edges to be welded vary from the specified pattern or location, the weld may drift from the edge producing a bad part. Vision would allow the robot to detect the seam and make adjustments to the movement of the welder to follow the seam precisely. Vision can also be used to watch the "puddle" and direct the speed of movement of the welder for optimum results. Some second generation robots are making their way into the market now with advancements inevitably to follow.

The third generation of robots will include artificial intelligence (AI) to allow the robot to react to conditions based on information gathered by sensors and interpreted according to rules programmed into the system.

Special-Purpose Robots

Not all robots are used for material handling tasks. In some notable cases, it is the tool that is being moved, not the work piece. The most common application of this principle is the robotic welder. Automotive assembly plants are famous for scenes of car bodies moving slowly down a conveyor while robot arms extend from both sides of the line, reaching over and into the shell, spot welding the pieces together in a shower of sparks.

Spot welding is a good robot application since the location of the welds can be predetermined and performed repeatedly. It is also a demanding task for a human — hot, physically difficult, and dangerous — and therefore a good use of automation.

Seam welding (typically arc welding) is a much more difficult application for robotics because it cannot be as precisely defined. Because of variability of the seam, the welder path is more difficult to define, exactly, ahead of time, especially with large work pieces. Also, the speed of movement should vary with formation and movement of the molten metal. Some robotic arc welders are now being developed with vision systems to monitor the seam location and the status of the "pool" of molten metal and to control the movement of the welder accordingly.

Another common application for specialized robotics, and a favorite of the auto industry, is spray painting. The spray head is mounted on the robot manipulator (end of the "hand") and can be precisely controlled as it moves in and around, through and under, back and forth around the sheet metal. Again, the use of robotics provides repeatable actions, reliable performance, and helps avoid human exposure to a hostile environment.

Some new welding and cutting techniques have been developed that rely, at least in part, on the ability to control the movement of the welding/cutting head over the work piece with specialized robotic movement controls. These special cutters use a number of interesting techniques to cut sheet metal of various gauges and often large sizes, as well as other materials such as ceramics, rubber, and paper.

As an example, high-power lasers can be used to cut steel and other materials with high precision and a clean edge. An advantage of laser cutting is the ability to cut formed (not flat) work pieces and to cut intricate patterns. Laser cutting is noncontact, and therefore does not place mechanical stresses on the piece and can cut varying thicknesses in one pass.

Other advanced techniques include plasma welding and cutting which use an ionized gas to conduct the current for noncontact arc welding/cutting (see Fig. 8-3). This yields a narrower effective tip than can be produced for traditional arc welding or cutting for enhanced results. Often, laser and plasma cutters are built into numerically controlled work stations that offer programmable cutting paths and rapid reprogramming capability, especially when connected to a program storage and downloading computer facility for distributed numeric control (DNC). The welder/cutter controls can resemble the controls found on a CNC mill, for example, with more limited movement capabilities than would be found on a robot-based design.

Work Cells

Extending the concept of an unattended machine, one that is usually associated with a robot or other loading/unloading accessories, to an automated process which may include more than one processing capability or facility, brings us to the concept of a manufacturing cell. Often a cell will include a fabrication process such as machining combined with another process such as drilling, and often a third step which could be an inspection station or an assembly machine.

Fig. 8-3. A robot being used for plasma cutting. (Courtesy Cincinnati Milacron Industrial Robot Division.)

There are an infinite number of configurations possible, made up of virtually any combination of machines and processes. What makes it a cell is the coordination of the processes such that the work is passed from one facility to the next in a continuous flow.

It should be noted, at this point, that a number of newer machines on the market are capable of multiple processes. There are multi-axis mills and lathes that can complete several processes on a work piece in one "step." A lathe, for instance, might perform the turning process, then hold the piece steady (not turning) while a drill and/or a milling head does its job. While the end result may be the same as that of a manufacturing center, the definition of a cell includes material handling (transfer) capability which, strictly speaking, these machines do not have. There is no reason at all, however, why a machine such as just discussed cannot be included in a manufacturing cell.

While manufacturing cells often include one or more robots to facilitate the transfer of the pieces from one facility to the next, this is not a requirement. It is perfectly feasible to arrange a cell with a human operator to load and unload the machines within the cell. Conveyor systems can also be used, or a combination of these (people, robots, and/or conveyors). Cells are customarily arranged in a "U" or circular configuration so that the operator or robot can easily reach all of the machines without unnecessary movement.

The chief advantage of a cell is the efficient movement of the work from one facility to the next, completing a cycle with minimal handling and delay resulting in short lead time and little work-in-process inventory (at least within the cell). From the discussion so far, it would be easy to conclude that the cell arrangement can be effective in almost any industry where the same activities (same parts produced) occur continuously, and that is true as far as we've gone. Single-purpose groupings of machines with mechanical transfer of work pieces are common in high-volume consumer goods industries.

The problem comes, however, when it is necessary to change over to another part or product. If the cell is very dependent on specialized tools, fixtures, and activities, then the changeover could take days or even weeks. Obviously, change in this circumstance would be undesirable, and long production runs would be required to be able to spread the changeover costs over as large a production quantity as possible. Today, in order to better respond to customer demand, we are trying to get away from inflexibility and the need for long (economic) production runs. We have come to understand that the high inventory levels that result from large-lot production are far more expensive than just their direct costs.

Inventory that becomes obsolete, no matter how cheap it was to produce, is money wasted. Long production runs and long changeover times also result in overall long lead times. If a single product is to be produced for two weeks to distribute the set-up costs, then other products that are made on the same facilities must wait two weeks for their "turn."

So, on the one hand, there are advantages to production cells such as short cycle time (lead time for a single part or product), low work-in-process inventory, and low production cost. On the other hand, expensive specialized cells tend to be suited only for high-volume continuous production and lead to lack of flexibility, long lead times, and high inventory. What is needed is the advantage of the cell approach without the restrictions, and that is just what programmable control and advances in automation have provided.

A Flexible Manufacturing Cell (FMC) is a group of machines, working together to perform a set of functions on a particular part or product, with the added capability of being conveniently changeable to other parts or products. An FMC is designed to consider the changeover requirements in the design of fixtures, capabilities, and programming. An FMC will normally be designed for a limited range of activities. These are usually targeted at a group of parts or products that exhibit some similarities in size, shape, materials, or processes required. The grouping of parts by their characteristics is called Group Technology (GT) and is discussed more fully later in this chapter.

An FMC is computer controlled, and the controller coordinates the activities between the machines and the transfer systems within the cell. The machines in the cell should have similar production rates so that each facility will be used efficiently. If there is a process in the cell that is significantly slower than the others, the slower process will determine the throughput of the cell and the extra capacity of the faster machine(s) will be wasted. A robot-equipped cell might have the work pieces moved directly from one machine to the next. An alternative arrangement would be to have the output from one machine directed into a temporary holding place (usually a chute or conveyor), called an accumulator, which feeds the next machine. The use of an accumulator has an advantage in that a short interruption at one machine does not necessarily shut down the entire cell, and machines with slightly different rates of production can be allowed to function at their "natural" speed somewhat independently of the other machines in the cell.

The design of a cell can be a complex and difficult task. It includes coordinating several processes with their individual production rates, designing the material handling equipment and (perhaps) programming the robotic movements, programming the cell controller to coordinate all of the cell elements, and arranging the cell components for most efficient use. With so many factors included in the task, it is often necessary to seek computer assistance in its accomplishment. Programs are available that describe the activities of cell components and assist in the development of instructions for robots and cell controllers. These simulation programs usually include graphic displays that allow the user to "see" how the cell will operate under various conditions and even in coordination with other cells and machines in the plant. Plant floor simulation was discussed in Chapter 5.

One key to success in the design and use of flexible cells is the adaptability of the fixtures and handling devices. Programmable machines (especially those with tool-change capability) and robots are easily reprogrammed, but the need to change fixtures can severely inhibit the flexibility of the cell. The use of a Group Technology approach to define the limits of size and shape of the work piece to be accommodated can be a major advantage in cell design and the resultant capability to change parts or products quickly with little or no human intervention.

The use of manufacturing cells changes some of the aspects of production planning and control. Obviously, the process description (routing) would reflect the use of the cell and its production rate. Activities (operations) before and after the cell should recognize the cell's input and output rate as well. If the cell is mechanically connected to other facilities through material handling systems, the rates of the other facilities and the transport systems should be coordinated to the cell's rate.

In today's Just-In-Time environment, we are less concerned with work center (machine or cell) utilization (hours of actual production versus hours available), but the cell will be most beneficial if kept busy as much as possible. Therefore, similar parts should be routed through the cell as much as possible (up to the limit of the cell's capacity) in order to get the best return on the investment. If there is more work than the cell can accommodate, however, there must be provision for performing the tasks at other facilities (internal or outside at job shops).

The ability to change an FMC from one part to another in a relatively short time is a major advantage, and impacts the planning of production and inventories.

Group Technology

The availability of the multifunction cell should be considered in the design and structuring of the part or product just as much as the part characteristics must be considered in the design of the cell. Once again, Group Technology is the primary tool used to identify those parts that have similar characteristics and therefore are candidates for similar processing at work cells.

Group Technology is simply the assignment of a code to each item (part or product) that serves to identify that item with others with the same or similar characteristics. There are many coding schemes in use, some commercially available and many more developed by a company for its specific needs. Groupings can be developed (assigned) for common characteristics of design, material, strength, size, shape, etc. A single item can carry multiple codes for different applications (design, production, purchasing, etc.) or multiple characteristics can be accommodated in a single, though more complex, coding scheme. There is little or no standardization in Group Technology coding at this time. A more complete discussion of GT was included in Chapter 5.

For use in conjunction with flexible manufacturing cells, the grouping code should reflect the required capabilities of the production process. These would typically include the processes needed (machining, drilling, grinding), perhaps the sequence of the processes, dimensional tolerances, physical size (to accommodate fixtures and robot grabber capabilities), machine speeds and feeds, and other production-related characteristics.

Group Technology codes would have to be assigned in the design and engineering process to enable process planning to use the codes to assign the part to the proper cell or process. As production processes are defined, cells assembled, and new products designed, the company should develop a plan for implementing a convenient way to communicate to process planning the characteristics of the part which would assist in developing the proper process description (routing).

Designed to be Made

From the foregoing discussion, you can infer that production considerations should be included in the design cycle for a new part. This concept is often called Design for Manufacturability or Early Manufacturing Involvement, and it is an illustration of the growing interdependency between departments within the manufacturing enterprise that is the driving force behind CIM.

Design engineers are problem solvers. They approach the job with the intention of finding the "best" design to satisfy the requirements. There can be many interpretations of the word "best," however.

Any given set of parameters (requirements) can be satisfied in a number of ways. There is a low-cost solution, a "cost is no object" solution, the prettiest, toughest, etc. It is the designer's job to come up with the optimum solution based on the requirements and other direction provided in terms of relative importance of these other parameters such as cost, ruggedness, etc. One of these other parameters should be production considerations.

It is not sufficient to say "Make it manufacturable," although in some companies this might be a startling new direction. Production, in communicating to design and engineering, must be specific as to what characteristics make the part easy or difficult to build. When there is extensive automation in the plant, the manufacturability parameters may be much more specific than in a manual situation.

Automated machines tend to rely heavily on jigs and fixtures that hold the part in proper position. Some shapes are easier to hold in a fixture than others. Robot "grabbers" (hands and fingers) have limited abilities compared to human hands, and this must be considered if the part is to be handled by a robot. Often, the addition of a small tab or flange, or the inclusion of a slot, chamfered edge, or keyway can make a tremendous difference in the ability to handle the part mechanically.

The following are some general ground rules that make designs more compatible with production techniques, particularly (although not exclusively) for automated production methods. While there are many different lists put together by automation vendors, consultants, and experienced users, the following illustrates the type of direction that the production people can give to design and engineering, and it is not intended to be an exhaustive treatment of the subject.

1. *Minimize the number of individual parts.* If two small pieces can be combined into one, inventory is reduced, handling is reduced, and one production step (at least) is eliminated. Expect the new part to cost more than the sum of the two "old" parts.

2. *Use existing parts where feasible.* Many times, an engineer will design a new part from scratch when an existing (off-the-shelf) part would be perfectly adequate. Existing parts will cost less and be more quickly and readily available. If already used on another product, they might even be in stock. Fewer stock items reduces inventory, more easily supports quantity discounts, and simplifies support (fewer spares and service parts).

3. *Avoid separate fasteners.* This one is particularly important for robotic assembly. Screws, bolts, nuts, and the like are hard to handle. Tabs, pins, and snap-fit connectors are preferred.

4. *Design for simple assembly.* Even if the plant is totally manual, some forethought into the assembly process can result in designs that require less labor and handling. Especially if designing for automated assembly, avoid twists, turns, and odd insertion angles.

5. *Design for multiple applications.* If the design can be such that it will be usable on multiple products, existing or future, then large savings can result from fewer inventory items, no need to redesign the component (reinvent the wheel), savings in tooling, machine programming, etc.

When applying any of these guidelines, remember that none is absolute. All factors are to be considered and weighed against the overall objectives of the project, and the best mix of characteristics will produce the optimum design. Most of these suggestions are self-evident common-sense ideas, and the main reason that they are not automatically applied in all cases is a lack of perspective. They have little or no impact on the design itself, and major impact on the production of the product. If design and manufacturing never speak to each other, how will design be able to consider production's constraints during the design process?

Flexible Manufacturing Systems

It is not unusual for a number of FMC's to be installed in the same plant. When they are connected together through a communications network, and materials are moved between cells with the help of material movement systems such as Automated Guided Vehicles (AGV's), they make up a Flexible Manufacturing System (FMS).

An FMS does not necessarily have to be made up of FMC's or only FMC's. The distinguishing characteristics of an FMS are 1) automated material handling between work stations, 2) centralized computer coordination/control, and 3) the ability to handle multiple parts (products) with minimal or no intervention required for change-over. An FMS may be a collection of automated machines (DNC), multiple flexible manufacturing cells, or a combination of these.

There must necessarily be some limitations to the flexibility; it is impossible to have one facility that can produce literally anything. An FMS is usually limited in the size of the pieces, the operations to be performed, and sometimes the sequence of operations. If the manufacturing capability has only limited ability to handle different parts, it might be more of a changeable transfer line rather than a true FMS. The ultimate in flexibility would be achieved with zero set-up time. Parts could then be run in any quantity, any sequence. While this is the ideal, the reality is minimal consideration of lot sizing and number of changeovers per shift.

An FMS starts with either DNC machines with tool change capability or flexible manufacturing cells or a combination of these. To the basic facilities is added a material handling system to move the parts from one work area to the next. If all parts to be run in the FMS follow the same sequence of operations, the material handling system can be relatively simple: conveyors, chutes, turntables. If a different sequence is required for different parts, programmable transfer facilities will be required (covered later in this section). Finally, the facilities are tied together through intercommunications and a controlling computer.

The interconnection is through a local area network (LAN) that carries the signal to and from each participant (node) of the system. There are several common configurations of these interconnections, called topologies, that describe the actual wiring paths. The most common topologies are star, ring, and bus (see Fig. 8-4).

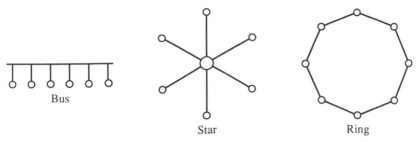

Fig. 8-4. Network topologies.

One prominent LAN on the factory floor is Manufacturing Automation Protocol (MAP), which was developed primarily through the participation of the "big three" in the Automotive Industry Action Group (AIAG). While there are a number of other networks in existence, MAP has the advantage of being an open system, that is, it is not a proprietary product and is available for use by anyone who wishes to comply with the network's requirements. A number of vendors of NC controls, computers, and communications equipment have MAP-compliant products available. In theory, at least, they should all be fully compatible with each other.

MAP uses a bus topology. MAP is also broadband, which means that coaxial cable is required, and uses a token passing protocol to move messages around the network. In a token passing protocol, each node is given a turn during which it can transmit on the network. Between turns, it can only receive, waiting for its next turn to respond. Other protocols have more random access which can be an advantage if the network traffic is light, but can delay (unpredictably) transmissions when the network is busy.

The cost to implement MAP is higher than some of the other protocols because of the more costly token-passing functions and the higher cost of coaxial cable (compared to twisted pair wiring). The user will have to weigh the higher initial costs against the more predictable response time, the advantages of using an open, standardized protocol, and the ready availability of equipment that is MAP-compliant. A more complete discussion of LAN's and communications protocols was included in Chapter 4.

The coordination and control of the FMS is the responsibility of a computer which is connected to all of the components of the FMS. The computer sends instructions to each of the facilities (machines, FMC's, material handling systems) and receives status information from each. Two-way communication is required because the computer must know what each resource is doing in order to coordinate the activities of the other resources to ensure proper processing and a smooth flow of work through the plant. The computer can also collect production information for historical reference and/or for use by other systems such as the business/planning systems and quality control.

Effective use of an FMS requires a good production scheduling system. In most cases, this means MRP. An MRP system (MRP II, see Chapter 5) is used to generate the requirements for parts at all levels and maintain the scheduling for production of all required items. MRP typically does not address individual machine scheduling nor does it concern itself with schedules more detailed than what should be run in a day. The finest resolution in MRP II is to the work center, which could be a group of machines that perform a certain task or, in the case of an FMC or FMS, MRP might only see a single facility which represents a "black box" view of an entire production process taken as a single production step.

Scheduling within the defined work center and within the individual tasks that make up the production step (as defined to the MRP system) are done either manually (supervisor's judgement) or within the plant-floor system. A fully integrated system would have the plant control system be able to receive the work center daily schedule information from MRP and develop individual machine or facility schedules and instructions, sequencing the jobs according to a priority, set-up considerations, and FMS capabilities. As long as the activities scheduled for the day are completed within that day, MRP is happy. The plant management system should be able to report back to MRP what was done, and the current status of orders not completed, so that an updated schedule and priorities can be generated and fed back to the plant system to restart the cycle.

Most MRP II systems schedule jobs through work centers based on a philosophy known as "infinite loading." The system has a defined work center capacity and daily production capability for a typical activity, and these are used to schedule each production requirement (job) individually without regard for the load that may be placed on the work center by other requirements in the same time frame. In other words, the MRP II scheduler does not consider work center load in the scheduling process — only capacity and rate. As a result, a work center that can handle three simultaneous jobs can actually be scheduled for five, or ten, or any number of jobs on the same day.

At first glance, this may seem ridiculous and unworkable but, in use, it is a reasonable approach that works well for many companies. Once the system has scheduled all of the activities, a capacity requirements planning function presents to the user a comparison of the scheduled load versus the capacity available (for each work center). The user can then decide how to handle any overload or underload situations by changing capacity or changing workload. The alternative to infinite loading is finite loading, in which the system levels the load to match the capacity available. This approach is difficult to accomplish and requires the system to make what amounts to management judgements.

When load exceeds capacity, there are a number of remedies: do nothing and miss the schedules; add capacity by buying equipment, hiring people, moving people around, authorizing overtime, adding another shift, etc.; or reduce load by shifting schedules, sending some of the work to an outside resource, canceling some jobs, buying instead of making, etc. The same kinds of options are available to handle underloads. With so many options, it is extremely difficult to design a piece of software that could make the judgements necessary to effectively manage these decisions. There are people in the industry who are very unhappy with infinite loading on a conceptual basis, but it is a workable technique that is being used effectively in literally thousands of companies. Several finite loading systems have been developed and marketed; some of them are large systems that are expensive and have often been difficult to implement.

Infinite loading is compatible with the integrated systems approach described here. Once the MRP-derived schedule and priorities are passed to the plant control system, detailed instructions are developed for the individual factory resources (sequence and schedules), and the results of operations are passed back to MRP II for update and replanning. As long as the load management (capacity requirements planning) is accomplished properly on the MRP II side, there should be no surprises at either end. Failure to adjust the load to the capacity, or the capacity to the load, will result in backlogs building up (in the case of an overload) which lengthens lead times and often causes missed schedules. In the case of an underload, resources may be underutilized. Idle machines were once to be avoided at all costs. Today, we recognize that utilization is not necessarily a good measurement of efficiency. This concept was discussed in Chapter 5 in the section dealing with Just-In-Time concepts.

When surveying your product lines and the parts that you produce that go into these products, it will probably be evident that a single, integrated, automated factory would have difficulty accommodating all of the various requirements for all of the parts. In the discussion about Group Technology and design for manufacturability, it was stated that production methods and constraints should be considered in the design process. Implied in that discussion was the fact that different kinds of parts have different characteristics, utilize different processes, and that it was desirable to categorize parts (using Group Technology) to direct them to the proper production facilities.

It may be the case that only a portion of your plant is organized as an FMC or FMS, and that other parts of the plant use more traditional methods. Or, you may have several separate FMS organized subfactories directed at different groups of parts. This subdividing of a plant into two or more subunits for specific products or production technologies is referred to as Focused Factories. Each focused factory is run fairly independently and is a complete production facility within its designed purpose, although it is not uncommon for one focused factory to supply parts to another within the same plant, or for several to supply parts to an assembly area (or focused factory). A focused factory is not necessarily an FMS nor is it necessarily flexible or automated. While the internal schedules and priorities within a focused factory are independent of what is happening in the rest of the plant, the overall production schedule (master schedule) must be tied to the plant's customer shipment or finished goods inventory objectives.

Material Handling

One of the concerns of a Just-In-Time effort is to reduce the handling and transport of materials and work-in-process as much as possible. Unnecessary handling adds to lead time, thereby increasing the WIP investment, and adds costs to the production process without adding value to the product — the JIT definition of waste. One result of a JIT effort is often a reorganization of the plant area itself, with machines and work centers relocated for more efficient material flow.

The traditional plant floor layout strategy (if there is any forethought at all) is to group facilities by function. Therefore, you will typically end up with a machining department, a drilling department, assembly area, inspection department, etc. While this might seem reasonable in terms of supervisory control, the flow of materials from one operation to the next often includes many trips back and forth across the factory floor.

The manufacturing cell and the focused factory are, in part, the result of a change in thinking concerning plant layout. Rather than group facilities by function, why not group them according to the "normal" flow of materials to minimize handling? The obvious problem that is self-evident at this point is that not all parts follow the same sequence. Many parts, however, are likely to follow similar sequences and have similar processes, and furthermore, parts can be easily grouped according to similarities in their processes — we are back to Group Technology and manufacturing cells again.

To accommodate groups of similar parts, most companies taking this approach will end up with a number of cells and/or subfactories, each of which produces a limited number of parts but does so with minimal handling. A bonus to this kind of reorganization is that the resultant plant usually requires much less space than the traditional departmental layout. Combine that with the reduced need for plant-floor storage space because of reduced WIP inventory, and the need for fewer tubs, carts, racks, and pallets, and some significant savings are available.

While, on the one hand, we have reduced the need for handling by concentrating the facilities needed to produce a given part into a smaller area, there is often a need for automated material handling within the cell, between cells, and between the plant floor and the warehouse.

Cell-Level Material Handling

There are basically two kinds of material movement systems that are used to transport materials from one station to the next: program controlled, and mechanical. Mechanical systems include all kinds of belts, chutes, and conveyors. These systems accept the work piece from the source (a human, a movement system, or a machine) and perform the same activity without regard for the product itself. Mechanical systems can be continuous, such as a belt which continues to move without any direct tie to the machine rate or cycle. Continuous systems in most cases require a buffer device at the delivery end, such as an accumulator which "stages" the parts for acceptance by the next machine.

Mechanical systems can also be synchronized to the process(es). These "indexed" systems will move the parts in increments which are timed to the rate at which the parts become available from the source machine. Indexed systems require that both the source and the destination facilities operate at the same cycle rate, or there must be a buffer (accumulator) at the destination end to accommodate different speeds or cycles. A typical indexed movement device is a carousel. Mechanical devices do not require programming and are not necessarily tied into a cell controller except perhaps for fault warnings.

Program controlled material handling refers to a robotic device. If the sequence of movement is fixed, the robot need not be automatically programmable through the cell controller. Many automated machines are available with robotic load and/or unload devices as accessories. While exhibiting the prime characteristics of a robot in terms of freedom of movement, these devices are not as sophisticated (or expensive) as fully programmable, flexible robots.

Many times, however, the requirements of an FMC include the ability to handle different parts and move them through different sequences. Programmable robots can provide this capability, with the appropriate program provided by or through the cell controller in coordination with the changes in the part being produced, and by reprogramming of the production machines. Both kinds of robots (special purpose and flexible) receive synchronizing information from either the cell controller or the facility (machine) from which the part is taken to initiate the movement activity. A synchronizing signal may also be required from the receiving device to let the robot know that it is ready for the next piece.

Between Cell Material Handling

Between cells, it is possible to apply the same kinds of technology discussed in the previous paragraphs. Instead of robots, however, the distances involved usually require the use of an Automated Guided Vehicle (AGV).

An AGV is a transportation device that resembles a cart or tow-motor-type vehicle that can operate through program control. The vehicle moves via electrical power (or sometimes compressed air) following paths defined by wires embedded in the floor. On command from the controlling computer, the vehicle will proceed to a predetermined location and wait for a handling device to load the materials. When the "load complete" signal is received, the vehicle moves along the prescribed path to the unload location where the materials are removed. It then proceeds to the next assigned location.

AGV's work best with standardized containers that can be easily loaded and unloaded by robots, conveyors, or rack systems with movement devices.

In a fully automated FMS, the AGV instructions can be generated automatically by the controlling computer to maintain the efficient flow of materials from warehouse to cell, cell to cell, and cell to warehouse. Multiple AGV's in a single system can be accommodated through a racetrack arrangement of the path (all vehicles travel in the same direction), although this requires a fixed route and can't minimize the "mileage" a part travels. A more flexible arrangement of paths with intersections can be managed if the vehicles have devices on board to detect other vehicles, and the central controller manages traffic and assigns priorities (handles conflicts). AGV's customarily play a tune or beep to let non-AGV's (people) know of their presence, and must be equipped with safety devices to prevent damage or injury due to collisions.

Warehouse Systems

On both ends of the material movement chain is the warehouse — for raw materials and components (either purchased or manufactured) and finished goods. Automated Storage and Retrieval Systems (AS/RS) assist in movement of items within the storage facility itself. These systems include dedicated facilities to move tubs, bins, and pallets into and out of rack-type storage areas. Some specialized AS/RS have been developed to handle individual items such as auto bumpers, but the vast majority of AS/RS depend on standardized containers.

The AS/RS control is combined with a location management system such that the system can identify the location of a particular item, lot number of an item, or group (quantity) of an item. The AS/RS may include the inventory accounting function or could be interfaced to a separate inventory system.

The system operator (or interfaced manufacturing control/inventory system) identifies the part to be retrieved, and the logic of the AS/RS will bring the appropriate item or container to a rallying point. There is typically a secondary material movement system associated with the AS/RS itself to move the items from the rack area to an access area where people, robots, or AGV's can receive them. This could be as simple as a conveyor belt or as sophisticated as an AGV.

While many AS/RS use moving vehicles of some type to retrieve the items from stationary locations and bring them to the marshalling area, some systems employ moving racks to bring the bin within reach of a less mobile retrieval device. These moving-rack systems are usually carousel or race track shaped devices which move a column of bins to a fixed point where a retrieval device which is capable of moving up and down removes the bin from the appropriate level (see Fig. 8-5).

The AS/RS has a sophisticated controller which is capable of translating a stock number, perhaps a lot number, and a quantity into instructions for retrieval and delivery. In a fully integrated system, this controller would be connected to both an inventory control system (either part of the warehouse system or separate) to keep track of quantities and locations, and also connected to the system responsible

Fig. 8-5. Allen Bradley Canada's Automated Storage and Retrieval System (AS/RS). When the baskets on the rotating carousel to the right are in position, the pod slides up or down to position the employee to receive parts. A light-tree tells which bin contains the needed parts while a computer terminal tells how many items are needed. (Courtesy Allen-Bradley Co., Inc.)

for tracking demand. This latter system could be the business system (MRP II) or the request could come directly from the plant floor.

Automatic material replenishment requests can be generated by an FMS or by a manufacturing cell based on scheduled activity, a plant- or cell-level stock management function (usually based on reorder point or minimum acceptable stock level), or a KANBAN-type system based on either usage of the materials or completion of the products.

AS/RS typically represent a considerable investment, and can only be justified in large-scale facilities that handle thousands of parts and thousands of material movements each day. The implementation of an AS/RS also indicates that there is considerable inventory to be handled, and this seems to be in conflict with the goals of modern inventory management philosophies such as Just-In-Time. If acquisition of materials from vendors and through your own production processes is timed to coincide with the need for those materials, then, at least in theory, there should be little or no inventory to store and retrieve.

Material handling in an efficiently run plant under the Just-In-Time philosophy would emphasize movement of parts and assemblies from one process to another, between manufacturing cells or departments, and on through the process to completion and out the door, rather than into and out of a storage facility. Just-In-Time also emphasizes reduction of the "mileage" that a part or product will travel during its processing to reduce handling and shorten lead time. These objectives coincide well with the development of manufacturing cells and the installation of material handling devices.

Summary

To summarize the organization of the automated plant, there are at least three levels within which automation is applied and communications systems exchange information. At the machine level, automated controls communicate with the machine itself, providing instructions according to a program that is executing in the controller. The machine may be equipped with sensors that will pass measurements and status information back to the controller.

On the second level, machines may be organized into cells in which the machines work in coordination to perform multiple operations on a work piece. The cell controller provides direction to the individual machines in the cell (through their controllers) and also to the material handling equipment that moves the work pieces from resource to resource in the cell. Communication between machine and cell controller normally uses direct connection, although a small LAN can be used. The machines feed production status information to the cell controller, and the cell controller provides instructions to the machines in the cell and accepts status information, synchronization signals (I'm ready for work. I'm finished — unload me), and problem/warning notices.

Cells may be grouped into a flexible manufacturing system or focused factory. Individual automated machines (DNC milling centers, assembly machines, etc.) may also participate as facilities within the FMS or focused factory. Communications between cells and the controlling computer are through a local area network using a defined protocol such as MAP. The controlling computer provides manufacturing instructions to the cells (or individual machines) in the form of schedules and sequences and receives status information back. The intercell material handling systems and AS/RS would also be participants in the plant-level network.

Fig. 8-6 illustrates the tiered structure of automation levels. The facilities at each level perform an independent function but also participate in an exchange of information with the levels immediately above and below. This arrangement exists entirely within the plant

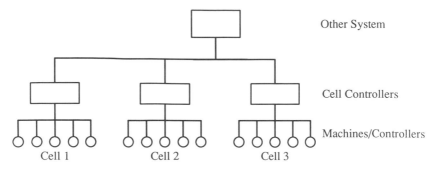

Fig. 8-6. Diagram of automated factor control levels.

floor area of the overall CIM diagram presented in Chapter 1 and repeated in Fig. 8-7.

Interfaces with the business and planning system would typically go through the plant's controlling computer either through a direct connection or via a gateway between the LAN and the business system. Often, there is a data collection system connected directly to the business system with terminals on the plant floor. The data collection system may communicate through the plant network or be independent.

The business and planning systems provide more high-level (lowest resolution is work center/day's priorities) schedule and priority information based on the master production schedules which reflect the corporate goals for customer shipment and finished goods inventories. The control computer refines the schedule and sends more detailed instructions to the various resources. Alternatively, the detailed scheduling (sequencing) can take place at the cell or machine level, except in an FMS. The control computer must feed status information back to the business system. The exchange of information at this level does not have to be interactive or real time. Since the plan is not ordinarily updated more than once per day, batch communication is usually sufficient, although it may be desirable to have on-line status information available at the business system level.

The interfaces with engineering can take place at any level. Machine control programs can be passed directly to machine controllers, through the cell controller, or via the plant LAN using a bridge or gateway between the engineering system and the plant LAN.

In a fully integrated system, the plant or the business system would notify the engineering system as to what programs will be needed by each facility, and when, allowing automatic downloading of the programs at the appropriate times. More typically, the programs will be held somewhere in the plant system (a file server on the plant LAN, the cell controller, or the machine controller itself), and only new programs and program changes need to be fed from engineering as they become available or when needed for production.

Chapter 8 Review Questions

1. What is a robot?

2. What is a manufacturing cell?

3. What are the advantages and disadvantages of cells?

4. What is DFM? What other names are used for this concept?

5. Why is DFM important?

6. Name several advantages of MAP.

7. What is infinite loading?

8. How can you resolve an overload situation?

9. What is a focused factory?

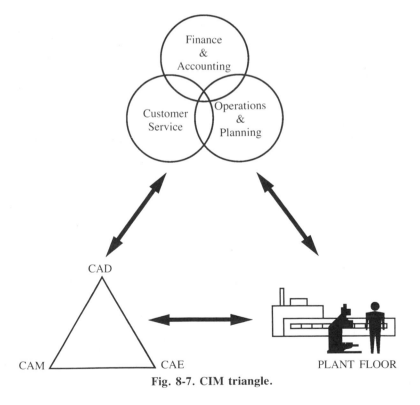

Fig. 8-7. CIM triangle.

The concern for quality starts with respect for your customer and pride in your products, and pervades all phases of activity throughout the enterprise. In recent years, with stronger competition and more demanding markets, the emphasis on quality has become an obsession, and rightly so.

Quality used to mean finding the "bad" parts or products before they were shipped out the door (hopefully), and either scrapping or repairing the offenders. Sometimes the dealers, distributors, and customers were unwillingly en-

9. Quality

listed in this final inspection process. When the customers said "no more," voting with their checkbooks, we got the message: if we don't deliver a quality product, someone else will. To stay in business, we have to improve quality.

Quality problems with parts and materials are also a major concern. If we can't assume that all parts and materials are usable, we must provide for extra supplies in order to avoid production schedule interruptions. Extra inventory costs money, must be stored, handled, and accounted. It sometimes is damaged in the handling and storage process, goes bad, or becomes obsolete.

Quality can take many forms. What comes to mind immediately is whether the part or product functions as it is supposed to, and whether there are any visible defects. These are the obvious measures. Beyond the obvious, however, there are many other measures of quality such as: is the product well suited to its intended use? Is it the most efficient design? Was it engineered to last? Does it provide real value to the customer?

The "less obvious" considerations listed above are not things that are easily measured and analyzed. They cannot be addressed by automated controls or statistical analysis techniques. They are inseparably tied to corporate direction, actual capabilities (including skills, technology, and resources), and personal attitudes. Generally speaking, these are subjective attributes, and are the focus of continuous improvement programs such as Just-In-Time and Total Quality Management, which are addressed at the end of this chapter and elsewhere in this book.

Direct measurement and analysis of physical characteristics of materials, parts, and products is the realm of Statistical Quality Control

(SQC) and Statistical Process Control (SPC). These quality management programs are used to identify and measure critical parameters, compare the measurements to the requirements (specifications), determine compliance or lack thereof, document the compliance rate, and provide historical information.

There are several major uses of these data. Statistical analysis of the foregoing provides information to be used in planning acquisition of parts and materials to be sure that adequate supplies are available to replace unusable items. This is the job of the planning system, and it is the user's responsibility to maintain the expected-loss information through maintenance of planning parameters so that production schedules are not disrupted by shortages.

Once the day-to-day concern of avoiding shortages is addressed, we can turn to problem analysis and improvement programs. Obviously, no one wants to make scrap. In order to reduce the occurrence of poor quality, it is necessary to identify the source and extent of the problem, formulate and implement a solution, then measure the results.

Statistical Process Control and Quality Control

It is a demonstrable fact that the earlier a problem is detected, the less costly it is to fix it. Field repair or replacement costs more than repairs before shipment. Repairs to the completed product cost more than repairs to assemblies or components. Preventing the problem is more cost-effective than fixing or throwing away bad parts. Taking this progression to the ultimate, it is better to control the process and prevent the manufacture of bad parts than it is to make the parts, inspect them, then determine which are good or bad. This is the concept behind process control.

Rather than wait until a production run is complete (or some other convenient sampling point) to count good and bad pieces, process control involves a continuous sampling and analysis aimed at detection of problems before any (or many) bad parts are produced. By monitoring and controlling the process as it occurs, the correction can be made at the earliest possible time. SPC (as well as SQC) uses basic statistical analysis techniques to characterize the measurements obtained and detect problems. The following section is an introduction to statistics.

BASIC STATISTICAL CONCEPTS

Statistics is nothing more than describing facts with numbers. Further, statistics assumes that things occur randomly, that is, a certain amount of variation is expected, and in most cases that variation is equally likely to occur above or below a midpoint of the range of measurements. This is called a "normal distribution" and is illustrated by a bell-shaped curve as in Fig. 9-1.

The graph shows how often each measurement occurs and illustrates that 1) small variations around the midpoint are more likely to occur than large variations (this is called central tendency), and 2) variations of a given amount above and below the midpoint are equally likely. While not all processes are exactly symmetrical, most are close enough to this ideal to make this model useful. For really skewed distributions, other models are available.

If the above statements are true, and if we can make enough measurements, we can figure out what the distribution looks like for the process being studied, and we can predict, very accurately, what future measurements might be and/or what the rest of the production lot would be like. The measurement of a sample, and the process of drawing conclusions about the entire group (population) based on the sample, is called statistical inference, and it is the basis for much of what statistics is all about.

Fig. 9-2 shows two normal distributions that differ in the width and height of the curve. They have the same midpoint, but the amount of variation exhibited by A is much smaller than that in B. We can say that A is more stable, more predictable, or more in control than B. Statisticians have measurements that describe the midpoint and the extent of the variability that can be used in quality and process control applications.

TWO BASIC MEASUREMENTS

The measurement of midpoint that we use in statistical analysis is called the mean — another name for the average with which you are no doubt already familiar. The mean is calculated by adding together all of the measurements and dividing by the number of measurements added. For any given sample, there is a figure which will have

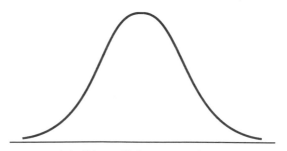

Fig. 9-1. "Normal" frequency distribution.

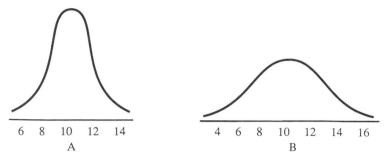

Fig. 9-2. Normal distributions with small and large variations.

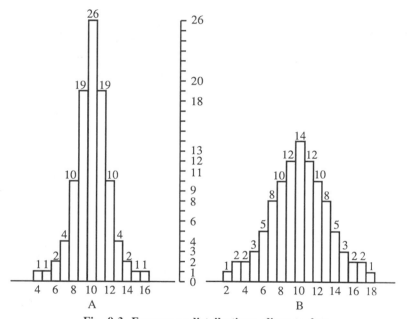

Fig. 9-3. Frequency distributions, discrete data.

exactly half of the measurements above, and half below. While the correct name for this midpoint is the median, in a normal distribution, the mean and the median will be the same or very close to one another. For simplicity, let's assume that the mean represents the midpoint. Half of the population is above the mean, half is below.

Fig. 9-3 shows the actual measurements used to produce the curves shown in Fig. 9-2. Both A and B have the same mean (10), however, in A, 84 of the 100 measurements were between 8 and 12. In B, this same range (8 to 12) includes only 58 of the 100 occurrences. You have to go out to the range of 6 to 14 to include 84 measurements in B.

If this were the length measurement of a fabricated part, for example, and the specification was 10 cm +/− 2 cm, the process in A produced 84 good and 16 bad out of 100, while B only produced 58 good and 42 out of tolerance. Obviously, process A would be preferable over process B.

Graphs are nice, but not always easy to obtain, and they are a little difficult to work with when using the data to make decisions. What we need is a number that describes the variability by characterizing the shape of the curve. This figure is the Standard Deviation, which is a measure of how far, plus and minus, from the midpoint you have to go to find a known percentage of the data. Statisticians use the Greek letter sigma (Σ).

A range of one standard deviation from the midpoint (plus and minus) includes a little over 68% of the data. One and one-half sigma encompasses nearly 87%. Going two standard deviations includes just under 95.5%. This relationship is illustrated in Fig. 9-4. There are published tables that include percentage values for increments of 0.01 sigma.

While the precise definition of sigma involves calculus, there is a simplified method for calculation that requires only squares and square roots. The formula is:

$$\Sigma = \sqrt{\frac{n\Sigma x^2 - (\Sigma x)^2}{n(n-1)}} .$$

The process is: square each of the data points and sum the squares, then multiply by n (number of points). Subtract from that the sum of the data points squared. Divide the result by n times $n-1$, and take the square root of the result.

You now can easily determine the range that will include whatever percentage of the samples that you wish. For example, if the mean is 10 and $\Sigma = 2$, 68% of the measurements will fall between 8 and 12. If the specification is 10 +/− 3, by using a table you could determine that 87% of the lot is expected to be within the tolerance (13% out-of-spec).

All of the above assumes that the process is "normal," and fortunately, this is often true. It also assumes that the sample is representative of the entire "population." Design of the sampling plan, tests for normalcy and accuracy of these inferences, is beyond the scope of the current discussion. If you are interested in studying this subject further, I highly recommend a basic course or seminar in statistics.

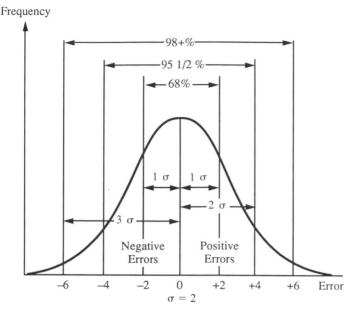

Fig. 9-4. The standard deviation measures variation.

A PROCESS MEASUREMENT EXAMPLE

Let's assume that a process is turning out parts that are measured (all or a sample taken continuously) to check conformance with a required length of 100 cm +/− 10 cm. The measurements yield a mean of 105 cm with a sigma of 5 cm. How are we doing?

As shown in Fig. 9-5, about 16% of the parts produced are outside of the specification.[1] This might have been acceptable in "the old days," but few customers will be satisfied with 16% rejects today. Companies now realize (as you, no doubt, do yourselves) that unusable parts are much more expensive than the cost of the parts themselves. There is incoming inspection, handling the rejects (repair, return or discard, report to vendor, claim credit, etc.), extra inventory to cover the losses, possible disruption of your production if sufficient inventory is not on hand, and expediting costs to get replacements in a hurry. Especially with more and more companies striving to operate on a Just-In-Time basis, reducing unnecessary waste and

---

[1]Since only one end of the distribution is of interest in this example, we use a probability table that assumes that the 50% of the samples below the average are all within spec. In this example, the table shows the low-end out-of-spec probability as less than two-tenths of one percent (less than two per thousand).

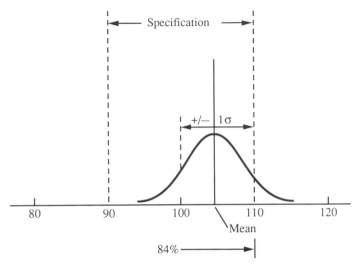

Fig. 9-5. Process example: mean = 105, sigma = 5.

buffers, anything less than zero rejects is becoming unacceptable virtually everywhere.

So, we have a process that is producing 16% bad parts. One option is to leave the process alone and perform 100% inspection, merely discarding or reworking the out-of-spec parts. Essentially, this is deciding that if the customer is unwilling to pay for bad parts, you will pay for them yourself. Another approach is to ask the customer to change the spec to meet what your process is producing. This is not the likeliest scenario, but is a possibility, sometimes.

The other approach is to change the process to produce parts that are within the specification. One solution is to reduce the variation in the parts produced. This produces a narrower curve, as shown in Fig. 9-6, which places more of the population within the acceptable range. For example, if sigma is reduced to 2 cm, the table shows us that 99.38% (2.5 sigma) will be less than 110 cm. We will have reduced the reject rate from 16 per hundred to about 6 per thousand.

Identifying the cause of the variation is not always a simple matter, however, and reducing variation is not always possible. Each process has a limit of its repeatability, and there is very little that can be done to tighten the process up beyond that limit.

Another solution is to change the midpoint. If we can lower the average length to 100 cm while maintaining the same variability (sigma = 5 cm), the reject rate will fall from 16 per hundred to less than 5 per hundred (see Fig. 9-7). This might be accomplished by

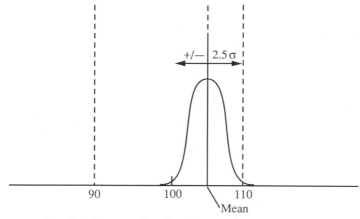

Fig. 9-6. Sigma reduced to 2 cm, average unchanged.

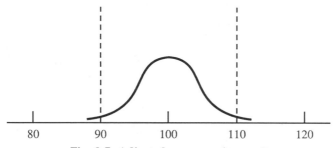

Fig. 9-7. Adjusted average, sigma = 5.

an adjustment to the process controls, adjustment to a jig or holder, or may require new tooling.

A combination of the above two solutions offers the best results. By bringing the average to the center of the specification range and reducing the variation to sigma = 2 cm, the projected reject rate falls to only a few parts per million. Centering the average and reducing variability to sigma = 4 cm, a more modest goal, results in expected rejects on the order of 12 per thousand (see Fig. 9-8).

As stated before, the science of statistics is based on the idea that a properly gathered sample can be used to represent all members of a group. The validity of the sample depends on the size of the sample and how it was taken. If, for instance, we take 10 samples from the first tray of parts each time the machine is started up, the sample may represent only the machine's operation during the first few minutes after start-up. As machine parts heat up from friction, the operator gets settled in, the jig seats itself, tool wear occurs, etc., the process (and the results) may change. It is important that samples are random and represent the full range of process characteristics.

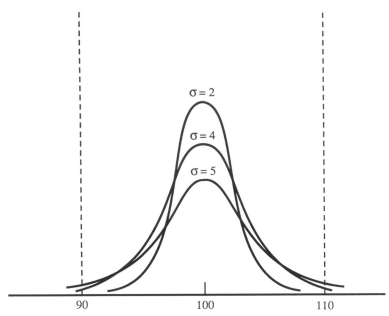

Fig. 9-8. Average = 100 cm, various sigmas.

Sample size and the resulting reliability of the data will depend on the history of the process. If the historical measurements are consistent and the variability doesn't change much, smaller samples can suffice. If the process is quite variable, more samples will be required to develop the same level of confidence in the statistics. Again, there are tables and formulas that describe the sample size requirements.

The ultimate sample is to measure the entire population. While this can sometimes be impractical or uneconomical, there are machines that will capture measurement information as items are produced and make it available to a computer or local network. Automated coordinate measuring machines (CMM's) can also pass measurement information on to an analysis application and are, more often, being included in automated manufacturing cells.

PROCESS CONTROL OR QUALITY CONTROL?

The concept of process control is oriented toward the identification of specific sources of production faults either as they happen or as process parameters approach out-of-boundary conditions, thus offering advance warning. Quality control is defined as analysis and maintenance of historical information that describes process capabilities. The difference is in the time scale more than anything: process control is identification and correction of problems in "real time," as the process is taking place; quality control is after the fact.

Both process and quality control utilize the basic statistical concepts and tools just introduced. In addition, SPC relies to a great extent on the use of several kinds of charts of process measurements to assist in identifying problems and monitoring conditions.

The most widely used chart for SPC is the Control Chart, also called the Shewhart Chart, named after the man who invented the technique (Walter Shewhart of Bell Labs) in the 1920's. The typical application of the Control Chart is a pair of graphs, as shown in Fig. 9-9, which track successive measurements of the average (top chart) and range (bottom) of samples taken from an on-going process. A common name for this chart pair is "X-bar R" chart, so named for the symbols used by statisticians for average and range.

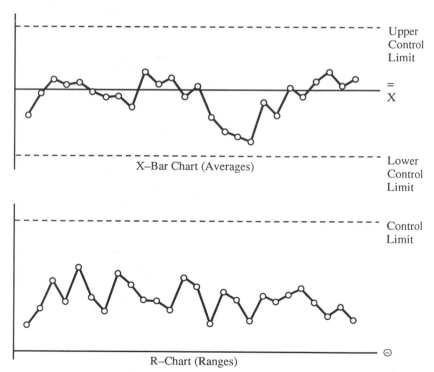

Fig. 9-9. Control chart.

To set up a control chart, first the process must be stabilized and measured. Once the mean and sigma are determined to be relatively stable, the charts are set up with "3-sigma" ranges indicated as "control" lines. This should include within the lines approximately 99.7% of the expected measurements. Samples from the process are taken at intervals, and the average of the sample group (usually 4 to 5 samples) is plotted on the top chart, while the difference between the high and low samples is plotted on the bottom chart. Twenty-five such plots are considered to be a "run" on a manually plotted chart.

As long as both graphs remain within the control lines, the process is considered to be "in control." An average that wanders out of range indicates a change in the measured parameter. A range that drifts toward the upper limit (lower limit for the range chart is zero) indicates increasing variation.

If the 3-sigma limits do not represent the required range as stated in the specifications, then the process is not appropriate for the requirement. Changes to the process to bring the 3-sigma limits within the spec are required. A 3-sigma limit represents about 3 rejects per thousand. If fewer faults are desired, control the process to the point where the specifications are beyond the 3-sigma level. Some contracts, in fact, specify the required variability in terms of sigma limits, typically requiring that the 3-sigma limit be at the specification range. Ford uses 4-sigma, and IBM currently uses 4.5-sigma for their vendors. Higher sigma limit requirements would result in fewer defective parts.

For automated monitoring, the process description (mean and sigma for the value being measured) must be determined using SQC techniques. Once the control limits are established, the automated measurement system can feed data to the application and the resulting control charts displayed on a monitor screen. In addition, the raw data and/or summaries or selected samples are stored or passed to another computer for storage.

The monitor application can also be set to perform an alarm function based on preset control limits. The alarm function can detect samples outside the limits and sound an alarm (audible as well as visual) at the monitor and also at a supervisory function located on the local network. The alarm condition could also be set to automatically stop production, change the sampling interval, and/or deflect the offending sample to a holding area or reject bin.

The alarm could also monitor a trend that, while still within the limits, could indicate an expectation of out-of-limit conditions in the near future. This anticipation of error supports a true "zero defects" approach by providing an alarm before bad parts are produced.

Problem Identification

In just about any production process that you can name, there are bound to be a number of possible causes that can contribute to any given result. That is, if a part is out of spec, any one or more of a number of things could have caused the problem. Once the problem is detected, it is necessary to identify the cause(s) and apply corrective action. Generally, a minority of the causes will account for the majority of the problems. This is the old "80–20 rule" or so-called Pareto analysis.

Identifying the cause of a problem, and keeping track of the occurrences of each problem type, are the key to understanding the process and what influences the results. With sufficient samples, the occurrences can be ranked in descending frequency on a Pareto diagram (Fig. 9-10). While it may seem obvious that the biggest cause should be addressed first, the selection of where to expend resources should be tempered by the impact of the problem; the number of occurrences may be high, but the dollar impact may be low. Also, any proposed solution must be cost-justified; is the solution worth the cost of the cure?

Any cost justification must consider the overall impact of quality, not just the unit cost of the pieces scrapped or rescued. Included are

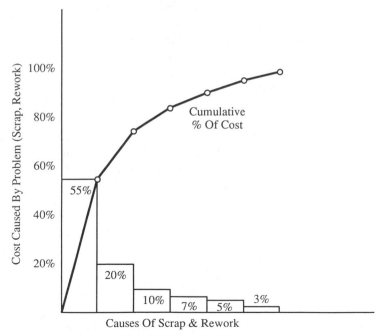

Fig. 9-10. Pareto diagram.

such factors as the cost of disruption and expediting due to unusable parts, lost sales or returns due to faulty products, delays in getting products to market, extra inventories to cover losses, etc. It is easy to overlook these factors when cost-justifying quality improvement efforts because they are somewhat indirect and difficult to quantify.

There was a very popular book published a few years ago that proclaimed that quality is free. The basic premise was that it is the lack of quality that is costly for reasons such as those outlined in the previous paragraph. To make an item correctly the first time is the basis of comparison, and all inspection efforts, rework, scrap, returns, etc., add cost. It is a logical conclusion, therefore, that doing it right is the low-cost alternative.

While this might make intuitive sense, if you are starting out with an imperfect process, there is bound to be some cost associated with improving the process to the point that only good parts are produced. It is these improvement costs that must be justified in terms of cost savings down the line in reducing or eliminating inspections, scrap, returns, etc.

Equipment and Systems

As mentioned, some controllers serve double-duty as data acquisition stations for measurements. The measurements can be displayed on a paper chart, stored in memory or on magnetic media (disk, diskette, or tape), or passed to other systems either directly or through a local area network. Direct data acquisition by or through a controller, and data from a coordinate measurement machine (CMM), would usually be continuous or 100% sampling. To maintain control charts, the data would be grouped and analyzed to obtain ranges and averages for the graph. The complete raw data, the group data, or selected samples from the raw data can be passed to the network.

Other data collection facilities can be used that are not a direct connection to the production equipment. Sample measurement using scales, calipers, and other devices with digital output capabilities can be used. In this case, the operator would select the sample, apply the measurement device, and press a "send" button to record and transmit the measurement.

Fully electronic systems are now being introduced which use vision systems to measure or verify part parameters. Vision systems use a video camera to capture an image of the object which is then sent to a processor for analysis. The processor must recognize the orientation of the object, measure the appropriate dimension(s), and compare the measurements to the specified limits. Image processing software has been very difficult to develop and requires a tremendous amount of computer power.

Other data collection devices that rely more on the human element include hand-held terminals for input of measurement or observation data and voice recognition systems. Data collection devices were discussed in Chapter 6. Inspection is a particularly good application for voice. A voice system leaves both hands free for handling and measuring and, if connected to an RF transmitter for wireless operation, allows freedom of movement for walking around large objects while recording observations. Bar-code systems are seldom used in quality data capture applications other than for identification of the item being inspected. It is easier to speak or type in observations (especially on a custom-designed hand-held terminal) than to use a bar-code wand on a template of defect types or descriptions.

Suitability for Use

In addition to the measurement of errors or nonconforming parts or products, the focus of a concern for quality must also be directed toward the design and suitability of the product for the intended use. The customer — the ultimate user — must perceive the product as suitable for its intended use, having sufficient value to justify the price, and superior to competitive offerings in some manner. Design and production of a superior product (by whatever measurement) involves the entire organization: *marketing* to identify a need and establish the general requirements; *design* to envision what the product would be like to fulfill the need; *engineering* to convert the design to specific parts and mechanisms; *process* or *industrial engineering* to define the processes and facilities necessary; *materials* to identify sources and coordinate acquisition of what's needed; *production* to carry through and actually make a quality product; *packaging, handling, shipping, distribution,* and *marketing* again to make the customer aware of the product's virtues; *sales* to generate the orders; *support and customer relations;* and *engineering* again for improvements.

This list is not all-inclusive; other participants might not be directly involved in the process, but are equally important in carrying out the work. For example: *maintenance* and *facilities* people are needed to keep things running properly: *finance* to manage the payments for all of this; *administration* to take care of the support services such as personnel, payroll, etc. As you can see, it is a team effort. Each participant must be focused on continuous improvement and the avoidance of errors in his or her own area of operation to ensure overall quality.

The concept of suitability and value applies not only to the customer but also to each successive internal "customer" in the chain. As each activity prepares its "product" for the next user, quality and suitability are important. Take a minute to read back through the list in the two paragraphs above. Each link in the chain of activities (and participants) adds something to what ultimately ends up as the final product. Errors and shortcomings at any stage will be carried forward into succeeding steps, with the result being a less-than-perfect product.

In addition, the same consideration is true in the parts and processes that are applied in production. Each carries its errors and imperfections into the next stage, the next assembly level, the final product. This applies not only to physical defects, but also to marginal compliance with specifications and lack of suitability or design shortcomings.

People

In quality, just as in the other areas we have discussed, it is the people issues that spell the difference between dramatic success and something less. While there are technological facilities that will help produce a better product the first time, and other systems that will monitor a process, automatically adjust the process for optimum results, measure quality, and provide analysis and alarm functions — if the operators don't care, if the company is not actively pursuing better quality in every way possible, it won't happen.

Many quality improvement programs include both training sessions for, and the active participation of, direct production employees, as well they should. One of the best ways to identify areas for possible improvement, opportunities for greater efficiencies, and advance warning of future trouble spots is to listen to the people who are out there, every day, in the plant, making things happen.

One of the most visible, and successful, efforts to tap this resource is the institution of "quality circles" within each of the key departments or areas of the plant. A quality circle is nothing more than an organized discussion session with the workers for the purpose of soliciting their ideas and enlisting their participation in the improvement process. Quality circles are only effective, however, if management is sincerely interested in the process and is willing to listen and act on the suggestions. Lack of follow-up will quickly dry up the flow of ideas.

Once again, attitudes will dictate how well this kind of program will work. The first attitude that has to change is the perception that direct production workers are not qualified to suggest process improvements. While the typical production worker may not be an industrial engineer, he or she may have an insight based on his or her participation in the process, at the source, that no IE can possibly learn in school or from a book. The opposite side of this relationship — the attitude in the workers that management is not interested in their input, or may react badly to suggestions — is equally of concern.

Suppliers

Because at least some of the parts and materials used in production are procured from outside sources, it is necessary to manage the quality of our vendors as well as we do our internal resources. In days past, the prime focus of the purchasing function was getting the best price. In an efficiently run company with a focus on quality, on-time delivery and supplier quality take precedence over price.

This is not to say that price is not important. Of course it is foolish to pay too much. But it is even more foolish, and possibly disastrous, to pay too little and not get the quality necessary. Just as your customers will buy value, you must also look at the total value rather than just the price of the part or material.

Late delivery or undershipment can cause severe disruption in the plant, especially if you have implemented MRP and/or JIT programs to reduce wasteful inventory. Inventory is a buffer against poor planning, changing conditions, and errors. If inventories are reduced, shortages will result unless the other factors are also reduced. Without the buffers, the errors, shortages, and nonperformance of vendors can bring production to a halt, cause changes in production schedules, require costly expediting, and/or cause you to miss *your* delivery schedules.

You must be willing to pay a fair price for your parts and materials. If the vendor must bid too low to "get the business," he/she will not make sufficient return to finance the production of a quality product, or will give priority to other more profitable work thereby delivering late to you, or might even go out of business leaving you with no source of supply.

It is important to work with your vendors, to provide them with enough planning information to assure capacity, to help them understand the requirements so that the product is suitable, and to help them to meet your needs. Good communications and a cooperative attitude are needed to counteract the traditional adversarial approach that has existed in vendor relations.

I used to take part in marketing oriented seminars in Baltimore. On several occasions, one of the major manufacturers in that area would come to these events along with a vendor. The purpose was to help these critical suppliers understand what MRP and JIT were all about and to understand their role in it. Hopefully, the vendor would at the very least come away with a greater appreciation of the importance of on-time delivery and 100% quality. Perhaps the vendor would be inspired to implement similar controls in his/her own production facility.

This "partnership" attitude of customer and vendor working to-gether is the only approach that will ensure reliable supplies — on time, and usable. Be willing to confer with your suppliers. Show them how and where their products are being used, ask their opinion. Vendors know their product and capabilities better than you do. They may have suggestions that can save you money, improve your product, or enhance their ability to deliver a quality product to you when you need it.

Many times, one result of the implementation of a Just-In-Time program is a reduction in the number of vendors used. This is not necessarily a requirement, but is a consequence of the amount of effort required to develop good vendors and the level of trust and interdependence that develops through this process. If a sole source or a limited number of vendors comes about, the vendor will benefit from more business from you (thereby having more at stake in keeping you happy), and you have fewer resources to manage so you can do a better job. Limiting the number of trading partners also simplifies and helps to justify the use of electronic communications such as are discussed in the next chapter.

Chapter 9 Review Questions

1. What do the mean and standard deviation measure?

2. If the average of 150 data points is 200 and sigma is 10, what specifications can be maintained at the 3-sigma limits?

3. How many bad parts per thousand would you expect in the above example?

4. What can you do if your process is producing out-of-spec parts?

5. What two things are plotted on a control chart? What is the typical group size used?

6. What is a CMM, and where are we likely to find one?

7. Is quality free?

8. What is a quality circle? What is management's role in the quality circle process?

9. Who is ultimately responsible for quality?

10. What are the most important considerations in vendor performance?

Every manufacturer is a distributor, to some extent. The manufactured goods must be sold and shipped to the customer whether that "customer" is the end user, a retailer, wholesaler, dealer, or a combination of these. All MRP II systems include the customer order handling functions as a matter of course, not only to complete coverage of this business area in itself, but also because customer demand as reflected in the flow of orders is a critical factor in developing master production schedules which drive the system.

10.
Distribution Functions and Electronic Data Interchange

In the simplest case, the finished goods are kept in a warehouse at the plant. When the customer order is received and processed through the order handling portion of the MRP II system, the goods can be packed, shipped, and invoiced. The invoice information is then passed internally to Accounts Receivable and General Ledger for further processing. The order information itself is a critical input to the planning functions as a statement of demand that is made available to material planning (customer demand in the master scheduling process), to forecasting (to develop a projection of future demand on which to base anticipatory production and purchasing activities), and to sales analysis to support marketing activities and the design process.

An extension of this situation could include multiple warehouses, all owned and/or controlled by the manufacturing company. In this case, finished goods are transferred from the production warehouse to the distribution warehouses from which they are sold. Each warehouse could control its own order processing and inventory control, they could be centrally managed, or a hybrid arrangement could allow local order handling with centralized planning and replenishment control.

More complexity is added when inventory is distributed on consignment to dealers or other remote facilities, when stocking distributors are involved, or when there is a combination of distribution resources and methods involved.

Assuming that manufactured items pass through a central warehouse function, at least logically,[1] the flow of goods out of the "production" warehouse can be used to forecast and plan total demand since all items sold out of other warehouses would have had to pass through the production warehouse immediately after manufacture. Finished goods inventory policy for the production warehouse is an integral part of the production planning and master scheduling processes.

The next issue is how to plan inventory levels at the other warehouses. What items should be stored where, and in what quantities? In this simple example, you can choose any of the available inventory techniques, including order point and MRP, or you might choose distribution requirements planning.

Distribution Requirements Planning (DRP)

In much the same way that MRP offers an alternative to order point for manufacturing components and finished goods, DRP is a planning philosophy for distribution that is based on planned needs rather than past usage.

With DRP, the focus is not on acquisition of materials to support production, but rather on how much inventory to keep at which locations to best satisfy expected customer demands. The distribution system can be set up in a two-level arrangement, with the primary (production) warehouse at the top feeding goods directly to distribution warehouses (Fig. 10-1A), or through one or more intermediate levels as illustrated in Fig. 10-1B.

With the multitiered arrangement, each warehouse places its demand on the primary servicing warehouse. In Fig. 10-1B, the distribution warehouse (DW) places its orders against the regional warehouse (RW) which, in turn, orders from the production warehouse.

Using the traditional order point management method, each layer of warehousing has its own stock and its own minimum quantity (order point and safety stock levels). As demand reduces the supply of an item at a DW, the order point is reached and a resupply is ordered from the RW. The RW ships the product to the DW reducing its supply and, eventually, triggering a replenishment order from the plant. All of this is going on at all the warehouses, with total demand

---

[1]In some industries, the completed items come off the production line directly into the trucks for delivery. While they do not physically move into "stock" then out again, within the management system it is usually necessary to reflect these two separate activities (work-in-process to stock, stock to shipping) in order to complete the tracking and accounting records.

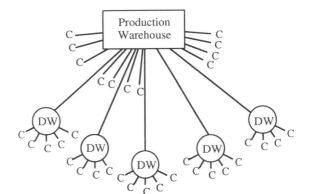

DW = Distribution Warehouse   C = Customer

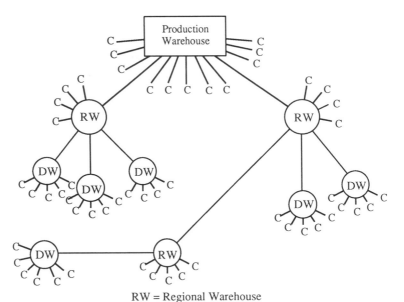

RW = Regional Warehouse

Fig. 10-1. A: Two level distribution system; B: multi-level.

ending up at the central plant. At each node, the order point is set at a level that provides enough remaining inventory to last through the replenishment cycle (the lead time to get more), so there are goods in the "pipeline" at all levels. This results in more inventory overall, which, in fact, is compensated for by reduced shipping costs (bulk shipments from plant to RW, and from RW to DW) and by reduced lead time to the customer because the warehouse is closer to him/her.

The biggest problem with order point, however, is that it doesn't look ahead. As demand increases, the inventory moves down the chain in a replacement scenario, with the total demand not reaching the plant until stocks have been transferred out to the ends of the distribution system. At that point, the plant must crank up production to renew its own supply and back-fill the RW's who in turn refill the DW's. Because all of this is in reaction to past demands (which, after all, determine the inventory level that is compared to the order point), resupply activity always tends to be tracking to past demands, not future needs.

Master scheduling at the plant must be based on forecasted demand. If the warehouse "pull" is passed down one or more layers by order point, the demand seen at the plant is offset from the real demand (delayed) as warehouse supplies are used up. Production, therefore, will lag behind demand and can become easily outstripped, especially in a seasonal or volatile market.

In the same way that MRP turns the orientation around in production from keeping up with past to preparing for the future (see Chapter 5), DRP bases replenishment activity in the distribution system on anticipated demands. The process is essentially the same. It starts with a forecast of demand. The expected shipments are subtracted from availability which is the current on-hand balance less any in-house orders not yet shipped, plus any expected receipts before the date of projected need, and an expected resulting balance is determined. This balance is compared to the safety stock requirement (if any), and if it is less than safety stock, a planned replenishment receipt is generated.

Notice that we are looking into the future and projecting when we will need to *receive* the resupply, not when to order. To determine the order date (or ship date from the supplying warehouse), subtract the lead time from warehouse to warehouse. If the planning is done far enough in advance (more than the required lead time), then the reorder can be on its way (on time) with a scheduled arrival date timed to the near-depletion of the DW stock (safety stock level).

The sum of the demands for each DW is used at the RW level, plus a forecast of any direct customer demands that the RW may be responsible for, and the same process takes place there. The resultant replenishment orders are passed along, time-phased for proper delivery, to the next higher level in the chain, be it another warehouse or the plant itself.

At the plant, we now have a forecast of customer demands modified by the ability of the distribution system to satisfy those needs through its "pipeline" inventories. We can use this forecast in the plant's master scheduling process to develop production plans, assure the availability of resources, and acquire the parts and materials necessary.

The biggest dependency of this whole process is sufficient lead time to make it all happen efficiently. If the lead time from warehouse to warehouse is one week, and there is one RW level in the chain, the forecast at the DW must be two weeks further out than the availability of finished goods at the plant warehouse. In other words, if a master schedule is developed at the plant to make inventory available on September 1st, the RW won't see the goods until September 8th, and the DW can't have them to ship to customers until the 15th.

Now consider the production cumulative lead time, including the time necessary to procure purchased items, and the total extent of the forecast becomes clear — manufacturing CMLT plus lead time through all levels of the distribution chain.

As with MRP, the use of DRP can result in much better availability and reduced inventory levels at the same time. The flow of goods through the distribution chain is managed to have the products in the right places at the right times, in concert with the expected demands. The chief dependency is on the accuracy of the forecast, which can be measured, and can be buffered with safety stock to cover some of the uncertainty in the forecasting process.

Electronic Data Interchange (EDI)

Automation and integration don't end at the front door of the plant. Since computers routinely handle purchase orders and customer orders, customer invoices and payments, it seems that the same move towards integration that is occurring within the enterprise can also occur between trading partners. Indeed, this is the case. The electronic transfer of such information comes under the general heading of Electronic Data Interchange (EDI).

The potential efficiencies available with EDI include significant cost savings due to reduced handling of paper, as well as considerable time savings in getting order information into the vendors' hands. While there appears to be sufficient justification for a move towards EDI on its own merits, in truth, few companies have openly embraced it because of a lack of standardization in the field.

As the first few companies experimented with EDI, they developed their own formats and protocols to handle the transactions that they were interested in — each company developing a custom approach geared to their immediate needs. A vendor, then, to participate with more than one trading partner, would be required to develop communications facilities and data-handling programs for each partner. As this could get quite expensive and unmanageable, there were few volunteers. The customers, however, started to insist that the vendors participate.

The move to EDI, therefore, has been driven primarily by major wholesale buyers of manufactured goods: the automakers and the large retail chains. In the late 1980's, the "big three" automotive companies and Sears, J C Penney, Wal-Mart, and others first encouraged major vendors to implement EDI, then later required EDI participation as a condition of doing business. The brute force approach has been hampered somewhat by the continuing difficulty in developing universal "neutral" standards, and the lack of data processing and data communications expertise at the vendor end of the line.

Virtually all of the "You must do EDI" ultimatums have either had to delay the implementation deadline or have had a significant number of partners avail themselves of exception clauses to secure one year (or more) delays. The conversion to electronic ordering and invoicing has been a slow, sometimes painful, process, but has proceeded nonetheless and now represents a significant portion of business in the automotive and retail chain store arenas.

The standards problem is still not completely resolved, but there is continued progress toward universal standards. The American National Standards Institute is developing a standard known as X12 for the U.S., and is cooperating with the International Organization for Standardization (ISO) efforts toward a world standard. Various industry groups have developed specific implementations of X12, such as the Voluntary Inter-Industry Communications Standard (VICS) for retail by Uniform Code Council (yes, the bar-code people) and the Auto Industry Action Group (the MAP/TOP organization) for their industry.

EDI Defined

EDI can be defined as the electronic exchange of business documents between trading partners, either directly or through a public network, and can include specific format documents such as purchase orders and invoices, or nonstandard documents such as free-form messages and data files. The formatted documents addressed by current and developing standards, in addition to the two listed above, include P.O. Change, Inventory Advice, Product Transfer and Resale Report, Price Sales Catalog, and Ship Notice/Manifest. In addition, the stan-

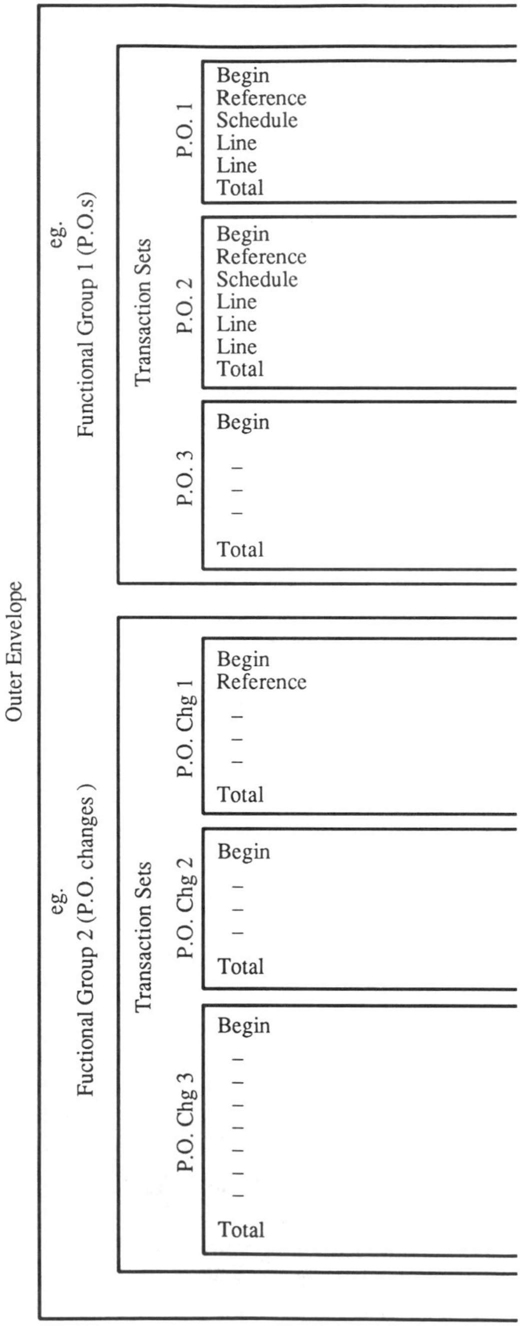

Outer Envelope

eg.
Functional Group 1 (P.O.s)

Transaction Sets

P.O. 1

Begin
Reference
Schedule
Line
Line
Total

P.O. 2

Begin
Reference
Schedule
Line
Line
Line
Total

P.O. 3

Begin
–
–
–
Total

eg.
Fuctional Group 2 (P.O. changes)

Transaction Sets

P.O. Chg 1

Begin
Reference
–
–
Total

P.O. Chg 2

Begin
–
–
Total

P.O. Chg 3

Begin
–
–
–
–
Total

Fig. 10-2. EDI hierarchy.

dards spell out how the transmissions are "packaged" with control records and verification procedures.

EDI communications can be direct — from the computer in my company to the computer in yours — or it can use a public network. Public networks are popular because they offer "mailbox" services which allow the sender and the receiver to each access the network on his/her own schedule and with his/her own communications parameters (speed, protocol). Some of the major public networks, also called value-added networks, at this time are Ordernet, Transnet, Redinet, and EDInet.

EDI addresses the application-level exchange of information and can be used with virtually any communications system or protocol. EDI definitions include the format of transaction data and how the systems identify each other and verify transmission and receipt. Transmissions are packaged in a layered or nested structure representing 1) the "outer envelope," which handles identification of the participants; 2) functional groups of like documents, such as P.O.'s, invoices, etc.; 3) within the functional groups are found individual transaction sets (documents); and 4) "segments," which are collections of defined data fields that make up the documents. Segments are like data records in a computer file and are subject to strict format definitions (see Fig. 10-2).

As an example, there is a transaction set (document) defined for a purchase order. A simple example of a transaction set for a P.O. might contain five different segment (record) types. Each segment begins with a two- or three-character identifier. Fig. 10-3 includes the entire EDI message according to the VICS definition. Discussion will start with the transaction set itself, and work outwards to the functional group and outer envelope.

The P.O. transaction set is contained in lines 4 through 14. The first record starts with BEG and identifies the beginning of a P.O. transaction set, and it includes, among other things, the P.O. number (24378) and date (870303). The second segment, the REF segment in line 5, identifies the ordering department (24). Lines 6, 7, and 8 are scheduling segments (SHH), and they contain instructions to ship complete on 5/8/87, cancel after 5/29/87, and don't ship before 5/1/87. The N1 segment (line 9) indicates that the ordering location is store 782.

The next four segments are the line items on the purchase order. Each identifies the quantity, unit of measure, unit price, and the Universal Product Code number. Line 14, beginning with CTT, is a transaction total segment that provides a check for the line item segments and serves as the end of the P.O. document.

1. ISA*00*bbbbbbbbbb*00*bbbbbbbbbb*08
    *61123450000bbbbb*08*6234560000bbbbb*870309
    *0817*U*00200*000000078*0*P*0;

2. GS*PO*6123450000*6234560000*870306*1300*1*X
    *002002VICS;

3. ST*850*0001;

4. BEG*00*RL*24378***870303;

5. REF*DP*24;

6. SHH*SC*010*870508;

7. SHH*P3*001*870529;

8. SHH*P1*037*870501;

9. N1*BY**92*782;

10. PO1**1*EA*21.95**UP*052177002578;

11. PO1**3*EA*21.95**UP*052177002585;

12. PO1**2*EA*21.95**UP*052177002615;

13. PO1**4*EA*21.95**UP*052177002653;

14. CTT*4*10;

15. SE*13*0001;

16. GE*1*1;

17. IEA*1*000000078;

Note: * in this figure indicates a data field delimiter,
        ; is used here to show the end of segment designater,
        b is a blank space (filler).

Fig. 10-3. Simple EDI purchase order.

Lines 3 and 15 in Fig. 10-3 are the transaction set control header and trailer segments. The header identifies that the set contains a purchase order (850). The trailer contains a segment count (13), as well as identifying this as the first transaction set within the group.

Lines 2 and 16 are the functional group header and trailer. The header (GS) identifies the type of documents in the group (P.O.), the sender's ID number, the receiver's ID, the date and time the functional group was created, a sequential number that is used to make sure that all functional groups are received and none is duplicated, a code that identifies the standard used (X = X12), and the version number of the standard used. The trailer (GE) contains a transaction-set count and a repeat of the sequential number from the header (group control number).

Lines 1 and 17 are the so-called outer envelope header and trailer. The header contains security information, sender and receiver ID's, date and time, interchange standard and version number, and finally a user-assigned control code. The trailer includes a functional group count and a repeat of the control code for checking completeness and catching duplications.

The structure of more complex transmissions follows the same pattern. The outer envelope segments contain a count of the groups, the group segments include a count of transaction sets, and each transaction set includes one type of document with the controls and counts for each set. Some of the segments are required, such as the envelope, group, and transaction set controls, while others are optional and are included only when needed. Some data fields within a defined segment type are mandatory, others can be omitted, but the delimiters must remain to show where a field has been skipped. The envelope control header is of a predetermined (fixed) length, and fields that are omitted or where the data do not fill the space must be filled to maintain the length. Other segment types are of variable lengths.

EDI software is available on virtually every computer type. The software provides a formatting capability, but a packaged general-purpose EDI program will not interact with your application. You must buy or develop a conversion facility to put your data in the proper format for use by the EDI package, and write or buy an interface to read the received EDI file and bring it into your application. Some vendors of purchasing and order entry systems have developed packaged EDI interfaces for their packages.

Sending the Message

Once the message is formatted with the proper packaging as illustrated above, it must be sent to the trading partner. One option is to communicate directly using a standard protocol, such as discussed in Chapter 4. The partners would agree on who would initiate the exchange, at what time, and under what conditions.

A more common scenario is to use one of the public networks with "mailbox" service to manage the exchange. Let's look at an example of how a group of purchase orders might be sent using a public network.

First, a schedule is agreed upon by the partners: the transmission to the vendor's mailbox will occur prior to a specified time.

The customer will use whatever planning and purchasing systems he/she has in place to identify the needs and generate the purchase order information. The P.O. data will be extracted from his/her purchasing system, formatted per the EDI standard being used, and packaged with the proper control segments and address information. During the agreed time interval, the message is sent to the vendor's mailbox in the public network.

At the specified time or later, the vendor accesses his/her mailbox and retrieves the packet. The file is interpreted by the vendor's EDI software, stripping the packaging away to reveal the P.O. segments in EDI format. A conversion program now reformats the P.O. data and passes them to the vendor's order entry system for processing.

The exchange is accomplished without producing or sending paper documents at the transmitting end, and without keying information into the order entry system at the receiving end. Both partners benefit through administrative savings, and the orders are in the vendor's system the same day they are sent. Eliminating the key-entry step also greatly reduces the opportunity to introduce errors. While most of us consider a reduction in paper as a definite plus, not everyone feels this way.

Legal Considerations

Since a purchase order is a legal document, a type of contract, and is issued and accepted subject to certain terms and conditions, there is traditionally an authorized buyer's signature and a statement of terms included on the document. In fact, commercial law (the Uniform Commercial Code) specifies that, in order to have an enforceable contract for the sale of goods for $500 or more, there must be a written document and a signature. With an EDI transmission, neither is included since there is no paper exchanged. The obligations of buyer and seller, therefore, must be specified and controlled by some other means, usually a "master contract" of some sort.

Among the serious considerations for such an agreement are issues such as what constitutes an order, what constitutes an acceptance, what proof is provided or required, and how are transactions verified. An electronic transmission contains data elements specifying quantities and item numbers, dates, prices, and shipping instructions. If there is an error on this order, how can it be detected and corrected? Who is legally bound in this situation if the goods are produced and shipped according to the faulty transmission? What remedies are available to either party? How can the EDI transmission be documented for valid proof for use in an audit?

How can the receipt of the order and its correctness be confirmed? If the vendor doesn't deliver, can you prove that the order was transmitted and received? When the goods arrive and the customer claims that he/she never ordered them, does the vendor have anything to fall back on?

These questions and many more, which remain unresolved at this time, provide the legal profession with expectations of considerable future business. Call the Mercedes dealer—this is the stuff that dreams are made of.

Relationships

The foregoing EDI discussion concerns customers and suppliers—and all manufacturers are both. The orientation is typically with ourselves as the supplier, since the demand for EDI has been forced down from the biggest customers—the auto industry and the large retail chains. As the technology improves and prices go down, and as standards become more stable and widespread, we may easily find ourselves with EDI connections on both ends—sending orders and receiving invoices with our suppliers as well as our customers.

When we seek out trading partners on the supplier end (whether for EDI or just in general), we will want to identify and develop those vendors who can treat us as well as we believe we treat *our* customers. The best way to put that into perspective is to stop and think about the things that our customers do (or could do) that can help us be a better supplier. For example, would we be able to provide better products, more effective designs, or better prices or delivery if we had more information about the ultimate use of our products? Would it help if the customer gave us some advance information about what his/her orders will be in the future (without obligation) so that we could use this in developing our master production schedule? Would we provide better service (quality and/or delivery) if the customer paid us a little more for each part or product?

Your attitude toward your suppliers is most visibly reflected in the measurements (reward and punishment environment) that you impose on your buyers. If the buyer is praised most often for securing the lowest price, or chastised for paying too much, the focus of your vendor relationship is based on purchase price. What is the reaction to quality (reject/return performance) and delivery (meeting lead times and/or due dates)? Is it as severe as the reaction to price performance?

In terms of the overall relationship, how willing are you to share planning and use information with key suppliers? The more you become a partner with a vendor, and the more information you are willing to share, the more you will be able to help your vendor perform to your needs.

Chapter 10 Review Questions

1. Why is order point troublesome in a distribution system?

2. Why is DRP like MRP?

3. What is driving the implementation of EDI?

4. What are some of the potential savings available through EDI?

5. What kinds of documents and transactions can EDI handle?

6. What are the hierarchical layers of an EDI message?

While not a part of the three defined areas of a manufacturing company that are the basis of the other chapters of this book, there are a number of office-type computer applications that are in common use and should be considered in any overall system integration plan. These include: word processing, general-purpose database management systems, Query functions, electronic mail, calendar management, spreadsheets, communications programs, and integrated combination packages.

11. Office and General Applications

Word Processing

Word processing was popularized by Wang and emulated by others, including IBM, with dedicated word processing systems such as IBM's DisplayWriter, until effective word processing software packages were developed for PC's. As in many other areas, the PC solution rapidly dominated the market and now represents the solution of choice in most environments.

While there are word processing packages available for midrange and mainframe systems, the PC is particularly well suited for this basically single-user application, and it benefits from the ability to transfer text files among systems on a local network or through the use of a larger system as a centralized storage resource.

There are a number of popular word processing packages in common use, and each employs proprietary procedures, unique use of keys, and incompatible file formats. The selection of a package depends as much on price, availability, and personal taste as on any technical feature or capability. Most companies with more than one word processing user will standardize on one package for ease of file transfer. Some word processors have the ability to import a file from one (or more) of the more popular products, and convert it to a usable format. Most packages can also produce and/or accept and convert an "ASCII text file" or "DOS text file."

ASCII refers to the character encoding method used internally in personal computers to represent the letters, numbers, and special characters as arrangements of "ones" and "zeros" inside the computer. It stands for "American Standard Code for Information Interchange," and is used by virtually all computer vendors, with the notable exception of IBM which uses EBCDIC (Extended Binary Coded Decimal Interchange Code) on everything but its PC's. DOS is the operating system used in IBM and compatible PC's. Both ASCII and DOS text files have the unique control codes of the word processing package removed. Importation of a DOS or ASCII text file into another word processing package often requires the reformatting of the document with new control codes such as "soft" carriage returns, type fonts controls, etc.

Some word processors contain extended features such as spell-checking, and the ability to combine a standard letter with a list of names, addresses, titles, etc., from a data file or data file extract. This function is called "mail merge" and is the key to generating "personalized" form letters.

Image Processing

One of the newest technologies to hit this area of computer usage is imaging. Introduced by Wang and quickly emulated by other major vendors, the ability to store, access, and display digitized images is seen as a leading-edge office application. The appeal is the ability to store handwritten notes in the original handwriting (sometimes useful for legal proof), pictures so as to verify damage estimates in an insurance claim, and other information that is not easily transferable to typewritten form.

Image technology is made possible by a new generation of digitizing equipment (scanners) and high-resolution monitors, along with the rapidly increasing capabilities and dropping prices of large-scale storage devices. Image is primarily an office application because it is a storage and retrieval function, not a processing one, and goes along with nonmathematical uses of computer technology such as word processing.

Database Management Systems

A basic "database" function allows a user to define a data storage format, enter information, and sort, retrieve, and manipulate the data without the need for writing programs. Database management packages have been a part of the PC scene for a number of years, with new and more powerful products being introduced every year.

The first and simplest packages had limited size and function, but provided significant capabilities especially in providing a nonprogrammer the ability to sort and search a large file of data. Many databases were used primarily to store customer and prospect data (for mailings and sales calls).

On the larger systems, database management functions were, and are, frequently included as a part of the operating system, or are sold as a system utility package, sometimes by the equipment manufacturer and sometimes by a third party vendor. One notable exception is the IBM System/38 and its replacement, the Application System/400 (AS/400),™ which have the database manager built into the design and structure of the system itself.

These database management systems (DBMS) are all proprietary and incompatible, although some can "import" data from other DBMS or spreadsheet systems. Each has its own unique file structure and procedures for storing and accessing the data.

A DBMS package will include a facility for accessing the information without writing programs. This facility is sometimes called a Query language or fourth generation language (4GL).[1] The state of the art in database design is the relational database management system (RDBMS). The relational model includes many sophisticated functions, such as being able to access multiple files as if they were one using a function called "join."

It is important to note that all data processing application systems include a database of some sort. The term database only refers to a collection of data stored in an organized way for a specific purpose. A DBMS is a utility system that includes the functions that manage the storage and retrieval of the data. An application (program or system) that is built onto a DBMS will use the facilities of the DBMS to manage the data, thus eliminating the need to perform these functions in the individual programs. A DBMS-based application system is more flexible, more secure, easier to modify and extend, and can be written and supported more efficiently than a non-DBMS application.

PC-based DBMS packages for general use (user-controlled data management and access as opposed to being primarily a programmer's tool) are often sold as a part of a combination offering which also includes other functions such as word processing, a spreadsheet,

---

[1]Programming languages have gone through several major phases or levels. Initially, programs were written as instructions directly usable by the processor (first generation). The second generation allowed the programmer to use a more human-understandable coding structure that could be interpreted by the system. The third generation includes the so-called higher-level languages—those that are converted to a machine-usable form through compilation. The fourth generation brings the coding process closer to a natural language (for the human), and is able to convert the instructions into functioning code and commands without writing programs according to strict programming rules.

communications software, etc. Integrated packages such as these have not been readily available on larger systems until recently.

Electronic Mail

Electronic mail, often called E-Mail, is an internal version of unstructured EDI. Typically limited to a single system, a local network, or a corporate environment, E-Mail provides the ability for the users to send and receive text messages. E-Mail provides the framework which includes access control (passwords), addressing, and mail-box services (store and forward).

E-Mail is most often used for free-format text messages, but some systems support more defined transactions. E-Mail also typically allows a message to be sent to any number of mail-boxes simultaneously and can store lists of committee members, department employees, etc., for ease of mass distribution. Simply enter the message, then call up the appropriate list of recipients, and send the message to all of them at once.

Many E-Mail systems also support "notes" which can be piggy-backed on top of a more extensive message so that you can let the recipient(s) know what is contained in the main message. Messages can also be copied, redirected, and easily responded to through system facilities.

Several large companies have successfully interconnected dissimilar E-Mail systems which were in use at different plants scattered around the country. Each plant had established its own network using whatever selected hardware and system they happened to use for data processing. To set up a corporate E-Mail system, the plants were linked using leased communications lines and gateway-type interconnect/converters to translate between protocols. Obviously it would be much easier to standardize worldwide, but it just proves that connections can be made to work despite basic system incompatibility.

Calendar Management

Many multifunction "office" packages contain calendar systems that display appointments and provide reminder messages to the users. If all employees maintain their calendars in the system, it becomes easy to schedule a meeting. The calendars of the designated attendees can be searched and compared until a common available time is located, then all can be scheduled into the meeting simultaneously.

Calendars can also be kept for facilities such as meeting rooms, overhead projectors, etc., and these can be scheduled and managed as easily as the people. When a user signs onto his/her office system, his/her calendar can be automatically displayed with reminders for the scheduled events for the day.

Calendar systems are seldom sold alone, but are typically included as a part of a comprehensive "office" package.

Spreadsheets

It has been said that it was the spreadsheet that fueled the popularity of the PC. Whether this is true or not, there's no denying that spreadsheets have become ubiquitous in both the office and the home. Most spreadsheets have traditionally been run on PC's, although midrange and mainframe versions have appeared recently from the major vendors.

Spreadsheets are, by nature, single-user applications, and therein lies the danger. These very useful tools are basically scratch-pad-type facilities that allow changes in data fields to be carried through automatically to other spreadsheet fields such as accumulations and totals. They lend themselves to easy manipulation, what-if analyses, and ad hoc updating. The problem is: each spreadsheet tends to be unique because of this updating.

Unless the user stores the spreadsheet in a central repository, and controls are in place to assure that the next user works from the same spreadsheet and puts it back into the repository when finished, there are apt to be many different versions in existence at the same time. You now have no integration and no control. If there are two or three (or more) answers to a question, which one is the right one?

Using a centralized data repository, or down-loading data from a centralized system into the spreadsheet, seems to be an obvious answer, but, in fact, once the data are loaded into the spreadsheet and manipulated, they will probably be stored (most likely on the PC) and now you have two versions: the original data on the central system and the spreadsheet version.

I have been an advocate of the "diskless" PC's in an environment with a centralized system since spreadsheets became popular. If the PC has no storage facility, the spreadsheet data must either be discarded after use or stored on the central system. While this is a rather unpopular position with spreadsheet users, the idea is beginning to catch on. In fact, there have been several recent announcements by major vendors of new diskless PC's specifically designed to provide PC power without the need, cost, or danger of local storage. These PC's are basically intelligent terminals that rely on the central system for storage of both data and programs, while removing some of the demand for processing power away from the central system down to the desktop. This is sometimes called cooperative processing — taking advantage of the relative strengths of each system type, such as the storage capacity and communications power of mainframes and midrange systems, as well as the local processing power of PC's.

Communications Programs

In the PC world, data communications capabilities for the masses were popularized by the Hayes Company with their "SmartModem" line. The dominance of this one company in the early years has established what has become a de facto standard such that nearly all PC modems are sold as "Hayes compatible." Hayes also developed a software program to manage the communications function (Smartcomm) making it easy for nontechnical users to dial the number, establish the connection, and send and receive messages and files.

Hayes' dominance in the communications software market is not on a par with their prominence in hardware. Other companies and educational institutions have developed a number of communications programs, some of them in the public domain, which perform these basic functions. Some of the well-known names are X-Modem, Kermit, and CrossTalk.

Early PC modems communicated at a speed of 300 baud (bytes or characters per second). The second generation upped the speed to 1200 baud. These modems encoded the signal (modulated on the transmit end and demodulated on the receiving end, thus the name MOdulator/DEModulator) using audio-frequency (tones). Twelve-hundred baud was about the practical limit for voice-quality phone lines. The quality of ordinary dial-up phone circuits prevents reliable communications at any higher rate. Recently, a new generation of PC modems has been introduced that operate at 2400 baud. Because of more sophisticated signal processing and error detection and correction capabilities, the effective speed of this kind of data communication has been doubled.

Even 2400 baud is agonizingly slow, however, for transfer of sizeable files and for interactive applications (two-way "conversation" with a computer program). Special lines can be leased which are of a higher quality (and, of course, a higher price) that are suitable for higher data speeds.

In larger systems, there are a number of proprietary protocols, with IBM setting the pace based on their large installed base. Many other vendors offer products compatible with IBM's standards, and there are a number of "neutral" communications standards that are in use throughout the country and the world. A discussion of communications protocols was included in Chapter 4.

Combination Packages

PC packages are commonly sold in combinations of the above functions. A typical offering might combine a spreadsheet, database, word processing, and communications. Combinations have become more common in larger systems, most notably in the midrange with word processing, database, calendar management, and E-Mail bundled together.

The main advantage of combination offerings like this, besides the attractive pricing they usually include, is the fact that they are integrated — designed to work together. The combination makes it easy to compose a letter in the word processor and "mail" it to a list of people with the E-Mail function, or combine the letter with an extract of the customer database to produce a "personalized" marketing letter.

Computers and Phone Systems

The in-house phone system, called PBX (private branch exchange), has also gone digital. Today's systems are computer controlled for call routing and advanced features such as call waiting, "park" (wherein a busy extension can be set to call you back when free), long distance routing (most economical method), and call accounting functions. New installations are also now installed with combination wiring that can also carry your computer network or terminal cable signals as well. With a broadband network cabling system (see Chapter 4), computer signals, telephone, video, and other signals can all share a single cabling system.

New technology is integrating computer functions with the telephone system. One that has gotten a lot of publicity, and garnered considerable controversy, is a service from the phone company that identifies to the call recipient (before the call is answered) the number of the calling phone. When connected to a marketing database, the person answering the phone can have, on his computer terminal, the customer's name and profile as he answers the call. This could be a real boon to customer relations, but people are concerned with their right to privacy and the potential for abuse. This capability, among others, is made possible by digital telephone facilities which are a part of the technology known as ISDN (Integrated Services Digital Network).

Phone-related software includes marketing support applications that are primarily database systems with the ability to dial phone numbers displayed on the terminal screen. This can be a great timesaver for telemarketing, and can be linked to word processing and mail merge to send out follow-up letters after the call.

The Fourth Area

In effect, then, we now have a fourth major area of automation to add to the other three that have been the focus of this book. Integration within the office area is quite advanced, and new products and services are being introduced as part of existing or newly formed integrated systems.

Because the office systems are most similar to the business and planning systems area, both by the location of the majority of the functions (front office, administration, and management), and by the kinds of system resources that are needed to support these functions (data management, many simultaneous users), it is not uncommon to find many of the office functions on the very same system (hardware) as the business and planning applications, or at least physically connected through PC networks and network gateway attachments on the midrange or mainframe system.

The office function that is most applicable to engineering and the plant is E-Mail. If these areas are "tied in" to the E-Mail system, interdepartmental communications is enhanced and paper and delays can be avoided. For example, customer demand information, design parameters, and marketing data collected or produced in the front office can be passed to engineering and/or production conveniently through E-Mail messages. Also, design changes and information about their expected impact can be E-Mailed to production control as soon as it is known, so that production can be prepared for the change release before it comes.

Of course, engineers and operations people can use word processing, spreadsheets, database systems, and the other "office" applications in their daily work.

Chapter 11 Review Questions

1. What is the most common location (platform) for word processing?

2. Name several uses for image systems.

3. What does the term database mean?

4. What is a Query function?

5. What is the strategic importance of E-Mail?

6. How can a calendar system save time?

7. Why does the author advocate diskless PC's?

8. What is the practical limit for dial-up communications speed?

9. What kind of cabling carries phone, data, video, and other signals simultaneously?

CIM, Computer Integrated Man-ufacturing, is not something that you can buy from your friendly local computer vendor or software company. In fact, there is even some uncertainty as to just what CIM is. The definition used in this book refers to the linking of data systems and functions throughout the manufacturing enterprise for the mutual benefit of all partici-pants. The benefits come not only from increased availability of in-formation, more quickly, and with fewer errors, but also in the better coordination of effort that results

12. Implementing CIM and Trends

from cooperation. CIM doesn't relate only to data exchange, but also to a recognition of all elements of the enterprise as interrelated, working toward a common goal, which is to produce a quality product, on time, at a cost that allows an acceptable profit.

You can, of course, buy equipment and programs that support the goal of more direct and effective exchange of information between systems. Many of the considerations for these interconnections have been discussed in the first eleven chapters. I don't expect to have made you an expert in systems integration in these few pages. I don't claim to know everything there is to know about the subject myself. There is a tremendous body of knowledge in data communications, systems integration, and the various disciplines found in a typical manufacturing company. Furthermore, new products and standards are being intro-duced every day. Hopefully you now have enough of a general under-standing to be able to deal with the hardware and software vendors and systems integration experts that you will undoubtedly come into contact with in your project.

As I said, CIM is not something you can go out and buy. At this stage of the technology, implementing CIM is a building process. Starting with the systems currently installed in the various departments of your company, you must identify the opportunities for integration, explore the technical challenges, and develop a plan to take you, step by step, from isolated systems to a smooth flow of information.

Recently, a new term has entered the vernacular: "Greenfield" CIM. This refers to the implementation of an integrated system right from the start in a brand-new plant. This is presented as the ultimate in CIM — designed to be integrated from day one. While few of us have as file building, parallel operation, learning new procedures, the inevitable "learning curve" — these represent an investment of sorts, to implement the new ways. This burden falls on the people who have the least time to spare and who typically had no involvement in the decision to make the change (this can be avoided!). These people will only participate wholeheartedly if it is made important to them. Clear, visible management support, commitment, and involvement in the project are the best ways to ensure the proper emphasis, to keep the project going when they are busy or when there are other demands on their time.

Commitment means more than a gratuitous statement at the first announcement meeting. Senior management (chief operating officer and functional area management: director of manufacturing, plant manager, director of engineering, etc.) must be *involved* and show an active interest.

I taught some initial classes for an MRP II project a few years ago and, at the first session, the program started with a video tape of the corporate CEO stating the direction and importance of the project. He clearly and eloquently stated that this was a major, strategic project for the company and that the company's very survival in the marketplace was dependent upon the success of this undertaking. I said to myself at the time, "This company will do well, the three elements are right here. We have strong management direction, the team is assembled here for training, and they are all committed."

About two years later, I happened to run into one of the team members at a conference, and I asked him how the project turned out. Despite the apparent good start, the whole effort stalled after a few months and failed completely to achieve its objectives. The team did their best, sufficient education was provided, but the management commitment ended with the video. When things got busy, there was no push to keep the project effort going. When tough decisions had to be made, the project lost priority because it didn't have the high-level backing it needed. A lot of money was spent. A lot of good people got frustrated and left. And the company was actually in worse shape than it had been before the start of the effort.

I hate to include such a negative example, but it is an important point which is best illustrated in the negative. Proper executive commitment, while an essential ingredient, doesn't necessarily stand out as a determining factor when looking at the results. By all means, take your share of pride and credit for a job well done, but recognize that the "owners" of the project are the real heros. Be sure that there is plenty of praise and recognition distributed to the project team and the entire company.

the opportunity to start from scratch, it is instructive to see how it can be done today with no existing restrictions. While we can't hope to emulate a Greenfield CIM site exactly, seeing the possibilities can provide insight to guide us in our plans to retro-fit CIM technology into our own existing plants.

You can hire (or rent) the expertise necessary to link dissimilar systems and pass information across previously impassable chasms. As you have seen in the previous chapters, there are a multitude of new products designed to make these linkages easier and more functional. Vendors of all types of systems are now making it easier to interconnect with other vendors' products, and "neutral" standards are developing which promise to make connectivity even more routine in the near future. The bigger challenge, however, may have nothing at all to do with data processing or data communications. Changing attitudes and encouraging a new, cooperative atmosphere among the employees in the disparate departments within the company is often a much more difficult undertaking than the technical connections discussed here.

I have tried to point out, in each chapter, the need to encourage a sharing of ideas, needs, constraints, priorities, and assistance throughout the enterprise. The impact of stronger cooperation can be much greater than the effect of technology changes. I'm not saying that technology can't help. There certainly is a need to enhance the flow of data throughout the company, but data alone cannot materially affect interdepartmental communications. Those supplying the data must understand how the recipients will use it, and the recipients must understand the source to appreciate and fully utilize what they get. This extends outside of the company as well. A vendor might well be able to provide better service and/or a better product if he understands fully how it will be used. We can apply the same reasoning to our customers. Product quality includes not only freedom from defects but also suitability for the intended task.

Planning for CIM

When Alice encountered the Cheshire Cat, she asked him which road to take. The cat asked Alice, in turn, where it was she wanted to go. When Alice told the cat that she wasn't exactly sure, the cat replied that it really didn't matter, then, which road she took. Like Alice, many companies really don't know what their destination is. They have no overall plan for integration, therefore, it's hard to tell if they are really making progress since there is no clear goal or path.

The planning process must start at the top with overall company goals outlining what the business will be like in two years, five years, and beyond. In line with general business planning, there should be some clear statements about the projected size of the company in dollar sales, units, employees, and facilities. The plan should include projections of the kinds of products and their volumes that will produce the sales dollars and how, in general, they will be produced, including the amount of subcontracting and the level of vertical integration anticipated. These estimates serve as targets for detailed planning of the resources and facilities that will be required to support the projected levels of production.

Next comes a realistic assessment of the state of technology currently in use. This is the easy part. Walk through the plant and list the kinds of machines in use and their capacities. Add to the list information about the automated controls, communications facilities, interface capabilities (whether in use or not), and the computing resources available and their communications capabilities.

Where possible, explore the options for expanding or adapting the interface capabilities and the cost, difficulty, and potential usefulness of these options.

Armed with this background information, start drawing a line from point A to point B. Project the changes in production and identify the need for additional resources to meet those needs as they grow. Speculate how system-to-system links could help tighten management control, make operations more effective, eliminate waste (anything that adds cost but doesn't add value to the product), and help support more business with a given level of the same resources.

Looking back at the nonsystem issues addressed in this book, make a list of areas in your company that can be improved through changes in procedures, different organization, education, discipline, and attention to detail. Put together a program to address these areas prior to, or along with, investing in any new technologies. Remember that any new equipment brought into the picture must be supported with adequate training for the users and an appropriate support structure (maintenance, continual training for skills enhancement and for replacements, new or revised procedures and controls).

Once you have developed a detailed plan, be sure that you put it in writing. The more formalized and detailed the written plan, the easier it will be to assess your progress, manage the project, and set priorities at each step of the way.

Even a written plan will not necessarily accomplish much unless each item on the plan has a name and a date assigned to it. Clearly defined responsibility, the proper authority to carry out the task, and

a real sense of "ownership" by the responsible individuals are the prerequisites for successful accomplishment of the project objectives.

The detailed plan must also have intermediate goals and measurement points. You cannot lay out a two-year plan, for example, and come back two years later to see if you made it or not. Regular progress checks (weekly, monthly) are needed to assess progress and provide further direction to keep things on track.

There will come a time in every project when there will be other priorities that jeopardize your schedule. Some tough decisions will be required; there will be some uncomfortable tradeoffs. One thing to avoid, if at all possible, is allowing the schedule to "slip." This sends a clear signal to the participants that the project is not as important as they were led to believe, and further conflicts and slips become inevitable.

When you set your original schedule, make it aggressive but feasible. A schedule that is too lax will encourage procrastination and will not achieve the improvements and payoffs as soon. A too-aggressive schedule will lead to frustration, protest, and probably failure. There's a fine line between too lax and too optimistic. The challenge is to find the right level and stick to it.

The participants, those who have their names next to the dates on the project plan, must agree with the objectives and schedules. Imposition of an aggressive schedule by executive fiat is not the way to instill a feeling of commitment and ownership. There should be room for some give and take in formulating the schedule, within the framework just discussed.

The Keys to Success

There are three things that, more than any others, determine the level of success that can be achieved in any CIM project, whether it involves the purchase and installation of new equipment or simply changes in procedures or organization. These three key elements have nothing to do with hardware or software selection, how advanced the technology, or how much money you spend. The keys to success are: senior management commitment, a team approach, and education.

MANAGEMENT COMMITMENT

Any improvement project will require some effort and some level of change, usually felt most acutely at the middle and lower levels of the organization. Often, extra effort is required up front; such things

THE TEAM

Integration projects, by definition, cut across functional areas. Often, they require formerly uncooperative or sometimes downright hostile groups to share information, cooperate fully for the good of the company, and consider each others' needs on a level with their own. An integration project cannot be carried out by one department and imposed on another. All affected areas must be full participants and take a share of the ownership and responsibility.

The project team should include the people who are the least able to take the time. The key players in the department or area, those who can "take charge" and "make things happen," are the ones who can make the project a success because they are the ones who are probably keeping their departments going. Everyone knows who these people are, and they earn the respect of their peers through their accomplishments. They are the leaders, even if not recognized as such by their job titles.

The project team is responsible for developing the schedule (with executive guidance and backing) and assigning responsible individuals and dates to each item. The team monitors progress, coordinates activity, and reports to senior management.

EDUCATION

The first job of the project team is to educate themselves. They must understand the need for the project, the "why" (theory), the effort required, the expected results, and the impact the changes will have on the organization, the procedures and activities, and the people.

Education and training must then be extended to all levels of the company that are affected by the changes. In most cases, the improvement project will involve changes in procedures, disciplines, day-to-day activities, and responsibilities. People have a natural tendency to resist change and fear the unknown. The best way to overcome these fears is through education.

Using Consultants

When dealing with new (to you, at least) technologies, it is often helpful to enlist the assistance of an expert in the field. A good consultant can help you understand and properly apply the technology, and can help you avoid costly errors by virtue of his experience with other installations.

A consultant should not, however, be your project manager, run the implementation team, or take responsibility for the project. You cannot delegate an integration or automation project outside of your company. Since the principal impact of these projects is on the culture and activities of the employees, the employees must retain ownership of the changes.

The consultant should be used as a resource for technical information, for the benefit of his experience, and for his intelligence and insight. He can also provide an objective viewpoint, not colored by your company's past experiences or current culture and personalities.

Some consultants will contract for extended services over a long period of time, such as one or two days per week for a year. While some companies and some consultants are comfortable with this kind of relationship, I have always preferred a lower level of involvement. A typical consulting arrangement for me would include no more than several days per month, and often only a few days per year.

I believe that there can definitely be too much of a good thing. If the consultant is to be on site and available every week, even if only for a day or two, I think the employees can become too dependent. Too often, the project doesn't really make much progress on the days that the consultant is not on site, and the employees don't develop the independence and self-reliance that is necessary to achieve permanent change.

During critical implementation phases, you may need assistance for several days or weeks at a time, but it is important to preserve your self-reliance. Don't think of the consultant as a company resource—he is strictly an advisor and cannot be held responsible for any portion of the project or company operations.

Another outside resource that you may utilize is contract technical services. Programmers and engineers can be contracted on a project basis to accomplish defined tasks such as writing interface programs, installing networks and cabling systems, or modifying equipment. These people or companies can and should be held responsible for their portion of the project. The difference is that the tasks assigned are one-time efforts that have a defined result and endpoint. The expected results can be clearly specified in the contract, and accomplishment can be proven through testing of performance of the product.

When contracting for services, be sure that there is a measurable product. Agree on the expected result and how it is to be proven to the satisfaction of both parties that the contract requirements are satisfied. Check references and be sure that the contractor has demonstrated his ability in the area that you will be contracting.

CIM Technology Directions

The trend in computer systems is up in capability and down in price, continuing in a direction that has prevailed for twenty years. The ubiquitous "personal" computer will grow more powerful each year, while last year's technology becomes a commodity product for sale at razor-thin margins by second-tier producers, often using significant foreign content.

Although the trade press has declared that mainframes and midrange systems have reached the end of their useful lives, like Mark Twain's famous observation, the reports of their deaths are premature. Vendors and customers have found larger systems to be very useful in a mixed environment, with PC's providing significant processing power at the desktop, reducing the demand on the centralized system which is focused on corporate data management, communications, and multiuser oriented tasks.

The emergence of the new smaller and more powerful workstations, somewhere between the PC and the traditional midrange, has somewhat confused market strategies and positioning in both markets. Large PC's compete directly with workstations in the scientific and CAD/CAM area while, at the same time, new business applications for the (mostly UNIX-based) workstations have the midrange vendors looking over their shoulders. In addition, several of the midrange vendors have recently introduced mainframe versions of their systems. These latter introductions indicate two things. First, the mainframe is not dead. If savvy companies like DEC are willing to make the investment necessary to go after this business, there must be significant business potential there. Second, the trend of smaller systems growing into market segments previously dominated by larger systems is a characteristic of the industry as a whole, not just PC's.

System connectivity is a major strategic direction for all systems in all markets. While the dominant vendors will continue to support and promote their proprietary protocols and standards, neutral, open alternatives will gain ground as will built-in or bolt-on-type adapters that allow easy connection to widely used communications links like Token Ring networks, Ethernet, and TCP/IP.

On the software side, more vendors will offer translation capabilities to neutral standards like IGES or the upcoming STEP standard. With more widespread development and acceptance of open protocols, information exchange should become more routine.

Proprietary operating systems will always be with us, but UNIX and its derivatives have shown surprising strength and longevity. Due, at least in part, to the increasing popularity of Reduced Instruction Set Computer (RISC) systems, more applications are becoming available for UNIX-like systems, and this segment of the market will continue to grow.

IBM's introduction in 1989 of its "CIM Architecture" and supporting products reflects a trend toward systemization of CIM. The tools and products encourage interconnection between individual resources, but the overall approach emphasizes some centralized controls and facilities as embodied in the CIM Architecture's repository (directory of data, its source, and usage) and data store (centralized storage function for shared information). Similar architectures and products can be expected from the other significant vendors of systems for manufacturers.

In services, systems integration contractors will continue to offer complete project management services (design, engineering, installation, etc.) for the larger companies. For smaller companies, the improved connectivity being built into new system products and the availability of "enablers," development tools that generate the interfacing programs with minimal technical knowledge required of the user, will bring do-it-yourself CIM connections within the reach of companies of all sizes.

Finally, continued development of neutral standards and the increasing popularity of open systems will bring more compatible equipment and software to the market from an increasing variety of vendors. While standards and open systems will never completely displace proprietary products, it is likely that only the biggest vendors will continue to develop and support proprietary systems and protocols, and the smaller vendors will comply with either the open versions, the biggest proprietary ones, or both. Even IBM now offers products that work with open systems and neutral protocols, although it has not, and most likely will not, abandon its long-standing policy of setting its own standards.

In Conclusion

Computer Integrated Manufacturing and the technologies that support it are developing at a geometric rate. The trend is toward more compatibility and connectivity, smaller yet more powerful and lower-cost equipment, and development tools that make CIM practical for even the smallest manufacturers. We live in an exciting time.

Answers to Chapter 1 Review Questions

Answers to Review Questions

1. The main distinction is that process manufacturing usually involves mixing, cooking, or a chemical process which doesn't involve simple unit quantities of components to produce units of the end product. Process industries are usually capital intensive (expensive, dedicated machinery involved), and production processes tend to be high volume with short production lead time. Bills of materials are more often referred to as formulas, plants tend to be more automated, keeping the production line "up" is usually a bigger concern than material availability or raw material inventory level.

While the label (process or discrete) is not in itself important, the characteristics outlined lead to very different emphasis in effective management of the plant. The automation needs, and facilities used, will vary with the importance of the various elements: process plants will worry more about production schedules (machine loads), process control, and perhaps finished goods inventory control. Much less emphasis will be placed on managing raw material inventory levels (shortages are a *large* concern), design engineering is replaced by chemical and/or physical research, and production schedules are usually shorter.

2. Business and planning, design and engineering, and the plant floor. They should be integrated because they each handle information that can be of beneficial use to the other areas.

3. Business and planning — engineering: product and process definitions. Engineering — plant floor: machine programs, process changes or difficulties. Engineering — business and planning: production instructions, activity reporting.

4. Plant floor: ability to handle many inputs and outputs, fast processing speed. Business: interactive (user) access, data management capabilities, storage. Engineering: mathematical calculation capabilities, raw processing power and speed.

5. Plant floor: microprocessors to programmable controllers and PC's. Business: mainframes to midrange systems, going to PC's. Engineering: mainframes to workstations, some PC's.

6. Interdepartmental rivalry and failure to redirect the motivational environment.

7. Increased worldwide competition, product specialization and niche markets (smaller production quantities), shortened product cycles.

8. MRP II is Manufacturing Resource Planning, the general name for automation of the business and planning functions with Material Requirements Planning (MRP) as a key application. JIT is Just-In-Time, a philosophy of continuous improvement and elimination of waste. JIT and MRP II are not competitive philosophies. MRP II is an approach to information management, and JIT is an operational philosophy that usually includes a reliance on information systems to support its objectives.

Answers to Chapter 2 Review Questions

1. 2D are two-dimensional drawings showing a single surface of the object with no representation of depth. 2½D uses a combination of two-dimensional information to represent depth in a view of the object. 3D is full description of the object in all three dimensions: height, width, and depth.

2. Identify the figure (line, circle), position, scale (size, length), and rotation.

3. Solids models can be developed in two ways. With boundary modeling, the edges and surfaces are defined using the techniques of 2D modeling. In constructive solids modeling, basic three-dimensional shapes are used in combinations to describe the object three dimensionally.

4. Primitives are the basic three-dimensional shapes used in constructive solids modeling. Examples are cube, sphere, cone.

5. The Opitz code consists of a group of five characters used to describe the form of the object. This group is most useful in retrieving drawings for reuse or modification. The second group of characters, four this time, describes the manufacturing parameters such as shape, original shape of the workpiece, and dimensions. This group is used to assist in process planning and manufacturing engineering.

The third group, optional and four characters, is for process parameters such as routing steps. Definitions for this group are unique to the user company.

6. While MIPS is a significant measure of internal processing speed, it is not the only factor in system throughput, the processing speed that the end user will experience. The only real test of throughput is side-by-side comparison with the same application.

7. Initial Graphics Exchange Standard (IGES) is an existing specification which covers only geometry and limited CAE capabilities. Product Data Exchange Standard (PDES) includes more production data but is not fully developed at this time. The future international standard is called STEP (Standard for Exchange of Product data) and is based on PDES.

8. The standard formats are still emerging and tend to change over time. Vendors are reluctant to develop products using a format that is likely to become obsolete in the near future.

Answers to Chapter 3 Review Questions

1. Computer Assisted Process Planning can be either through a *retrieval method*, under which existing routings are retrieved and modified, or *generative*, in which the process is developed without referring to existing routings.

2. Group Technology is a common method used in identifying similar parts or processes for the retrieval method of CAPP. Group Technology is also used to identify manufacturing cell capabilities and characteristics to help in process planning.

3. Each manufacturing facility has different capabilities associated with the facilities, personnel, and skills that are available to production.

4. Yes. Proper selection from among several machines or other facilities that are to be used to perform a function can depend on the available capacity of the facilities and planned or existing demand (load).

5. Various postprocessors are sold to convert the tool path model into NC instructions for different machines.

6. Robots can be programmed directly in the same manner as other PLC-controlled devices specifying motion/position for each controllable device (joint). This is the most difficult method.

Many robots have a "recording" system where successive positions are recorded and the controller develops the motion commands. This is a much easier method if CAD-based program development software (the third and best method) is not available.

7. Perhaps, but don't count on it. The program should be verified on the production machine or robot, and corrections made if necessary.

8. 1) Enter the bill information in the CAD system in level-oriented manner and store in the CAD file. 2) Extract from the CAD file and convert to the file transfer format. 3) Transfer. 4) Optionally convert to the business system format. 5) Read-in to business system and analyze for differences and/or unknown items.

9. A "surprise" engineering change can obsolete parts in inventory or on order; can make current production activity inappropriate, causing rework or scrap; can cause expediting of parts that are new or different and are needed right away; and doesn't allow time to prepare production facilities or processes.

Answers to Chapter 4 Review Questions

1. 1) Physical, 2) data link, 3) network, 4) transport, 5) session, 6) presentation, and 7) application.

2. LAN products usually address layers 1 and 2 only.

3. OSI can be applied on virtually any LAN type including Ethernet, Token Ring, broadband, and public networks.

4. Ethernet is a baseband, CSMA/CD nondeterministic network.

5. A frame is a bundle of data bits identified by a header and trailer that define its beginning and end. This is the basic unit of transmission at layer 2. A packet is the unit of transmission (bundle of data) that is handled by layer 3 and does not contain the layer 2 header and trailer.

6. Carrier Sense Multiple Access/Collision Detect is a nondeterministic traffic control strategy. As the network gets busier, response time increases. Token-based systems are deterministic, which means that there is no significant change in response time as network traffic increases.

7. MAP is a broadband token-bus network.

Answers to Chapter 5 Review Questions

1. Order Point assumes shortages (service level) which can only be compensated through extra inventory (safety stock). Order Point looks back, not forward, thereby not recognizing changes in demand often leading to shortages and/or excess inventory. Order Point doesn't work well for dependent components since probabilities multiply.

2. MRP requires accurate bill-of-material information, accurate inventory records including on-hand balance and availability, lead time estimates, and a master production schedule.

3. MRP recommends the release of new orders, recommends changes to existing orders, and provides priority information.

4. The shop floor must provide order status information to the business and planning system. Typically this includes reports of activities performed, the time taken, and quantity produced and scrapped.

5. Theoretically, MRP can result in no inventories and no shortages, but it usually doesn't because a) there are inaccuracies in the bills of material and inventory records for which we compensate with inventory, b) lead times are often overstated resulting in delivery of parts before the actual need date, and c) lot-sizing considerations often result in production lots and purchase order sizes that do not correspond exactly to the needed quantity.

6. KANBAN is an inventory and production control technique that uses cards (or equivalent) to signal replenishment activity.

7. With MRP II and/or JIT, in an effort to more closely coordinate all activities toward production goals, we must be more concerned with vendor performance than strictly purchase price. We must develop vendors who will deliver on time and who will meet our quality objectives reliably. We may also be willing to develop closer, more cooperative relationships with vendors and be willing to share planning information with them to help them serve us better.

8. Just-In-Time is a management philosophy that emphasizes continuous improvement. It is a continuing campaign to identify those practices that add cost without adding value (defined as waste) and reduce or eliminate this waste. It is not a piece of software or any specific technique, but often uses MRP II and various other tools.

Answers to Chapter 6 Review Questions

1. Manufacturing: Code 39. Retail: UPC. Government: Code 39. Cartons: Interleaved 2 of 5.

2. Code 39: one character which is both start and stop. UPC: has no start or stop characters but uses guard bars at the beginning, end, and in the middle of the symbol. Interleaved 2 of 5: one each. Code 128: three start characters (for the three character set — A for upper case, B for lower case, and C for numerics).

3. Code 39 has the best, Interleaved 2 of 5 has the worst.

4. OCR has high error rates, and is harder to scan because the scanner must be precisely aligned with the code characters. Magnetic strips are expensive (relative to bar-code printing) and cannot be read at a distance in a noncontact mode.

5. Accuracy will improve to the extent that errors introduced through the reporting and handling process itself will be greatly reduced. These include such things as poor handwriting, miskeying, lost "tickets," number transpositions, etc. The data collection system itself will not change the interest or motivation of the users outside of any "Hawthorn Effect" that might be present.

6. An event can be the beginning or end of an activity. A transaction represents the results of an activity often involving a pair of events as defined above. An event happens at a moment in time, a transaction might include an elapsed time calculation.

7. The user does not have hands free for reporting. Writing or wanding takes too long. User is wearing gloves, has messy hands, or is physically disabled.

8. Data collection doesn't make chips (does not produce parts), therefore, the justification is mostly indirect.

9. Data acquisition is the automatic recording of process parameters on a paper chart or electronically (memory, disk, tape, or passed to another system through a communications line).

Answers to Chapter 7 Review Questions

1. Closed-loop refers to the availability of sensor-derived information fed into the controller which is used in the control of the process.

2. Ladder logic is still in use primarily because of inertia. The incorporation of ladder logic in early controllers significantly contributed to the acceptance of the new technology by providing a familiar interface. Ladder logic is not well suited to feedback control applications (as opposed to sequencing) and could be replaced by more effective methods.

3. A Personal Computer (PC) is designed for general data processing tasks and is limited in its ability to handle many inputs and outputs. A Programmable Logic Controller (PLC) is designed to handle many signals in and out, but tends to be less well suited to data management, storage, and user-interaction requirements. Technological developments have blurred these distinctions, and PC's and PLC's can be found side by side performing many of the same tasks.

4. PLC's can be programmed by direct entry through the operator station, by a recording technique, through loading a program previously stored on tape, or by downloading from a computer.

5. DNC stands for either Direct Numerical Control or Distributed Numerical Control. In DNC, controller programs are stored on a computer and downloaded to the controller over a communications line.

6. Engineering provides the bill-of-material information and process instructions (routing, drawings, specifications) to the plant floor. Business and planning provides directions (what to make and when) and priorities.

7. The plant floor provides activity reports to the business system (order number, routing step, time spent, quantity good and scrap). The plant must keep engineering informed of any errors or recommended changes to bills and process information.

8. Unit cost can be reduced by 1) reducing set-up (fixed) costs, 2) reducing overhead, 3) reducing process variable costs such as run time, and 4) through larger lot sizes (distributing fixed costs over a larger quantity).

9. The biggest impediment to effective implementation of management systems is the "people" issue. All personnel must be included in the planning and implementation process, all must be informed as to the reasons for the change and the expected impact, and all must be properly motivated to actively support the new way of doing business.

10. It is better to allow a machine to sit idle than to produce parts that you may not need. An idle machine wastes space and fixed costs, but producing useless parts also wastes materials, labor, storage space, etc.

Answers to Chapter 8 Review Questions

1. A robot is a programmable machine whose chief characteristic is its arm-like structure and movements. Robots are used mainly for movement and placing of objects or specialized tasks such as welding.

2. A manufacturing cell is a group of two or more manufacturing resources (machines) that are placed in close proximity and operate as a unit to perform multiple operations in coordination with each other. Cells sometimes include a robot for movement of the materials from one resource to the next.

3. The advantages of a cell include short lead time for multiple process steps, reduced handling, and less "mileage" for the materials or parts. Disadvantages include limited function, as cells are specialized for specific tasks and sequence and the associated scheduling difficulties.

4. DFM stands for Design For Manufacturability, and refers to the concept of a designer consulting with production engineers to make sure that the design of the part or product is appropriate for manufacture. Other terms are Early Manufacturing Involvement, Simultaneous Engineering, and Concurrent Engineering.

5. DFM and early involvement programs shorten the design-to-delivery time, and usually result in a better product that can be manufactured more efficiently and can better use the resources available in the plant.

6. MAP is an open standard which means it is not specific to one vendor, it is broadband which supports multiple use of the network's physical resources, and it is token-passing which provides relatively constant response time regardless of network traffic conditions.

7. Infinite loading is a scheduling philosophy in which each job is scheduled according to work center capacity without regard to other demands on those resources during the same period. Mismatches between load and capacity are resolved by production management based on information available from a capacity requirements planning application.

8. An overload at a work center can be resolved by increasing capacity (adding people, adding machines, authorizing overtime, adding a shift) or by reducing load (rescheduling jobs, sending work outside for processing, canceling orders, using alternate resources).

9. A focused factory is a portion of the production facilities that has been separated out and organized for efficient production of a limited set of tasks (products), and is operated independently of other portions of the plant.

Answers to Chapter 9 Review Questions

1. The mean measures the midpoint of a distribution, and the standard deviation measures the variability.

2. Specifications of $200 +/- 30$ can be maintained (170 to 230).

3. Approximately three bad parts per thousand.

4. The alternatives are: ship bad parts, inspect 100% and discard the bad parts, measure and analyze the process and either reduce the variability or center the distribution at the center of the specified range, or (best) both.

5. Control charts include a plot of the group averages and the group ranges. A group is typically 4 or 5 measurements.

6. CMM stands for Coordinate Measurement Machine, a device for measuring physical size and shape. CMM's are often found in manufacturing cells, but also reside in QC labs.

7. In theory, making only "good" parts is the most cost-effective alternative since inspection, rework, scrap, etc., all add cost. Starting with an imperfect process, there is usually some cost associated with improving the process to increase quality.

8. A quality circle is an organized discussion group of direct production employees set up for the purpose of soliciting their suggestions for improvements in quality and efficiency. Management must be visibly committed to the program, must be willing to listen to the suggestions with a completely open mind, and must follow up with implementation of the best ideas.

9. Everyone.

10. On-time delivery and quality are the most important factors in vendor performance. Price is a secondary consideration.

Answers to Chapter 10 Review Questions

1. Order point results in a large inventory investment and questionable customer service (high potential for shortages). In addition, changes in demands are not foreseen and the plant could be unable to keep up with increases in sales.

2. Like MRP, DRP relies on a forecast and a calculation of what is needed when, with replenishment activity timed to bring in the needed resupply only as the stock is about to run out, not to satisfy a specified minimum stock level.

3. Participation in EDI is being demanded of suppliers by major customers such as the auto industry and large retail chains as a condition for doing business.

4. EDI can greatly reduce the handling of paper, can eliminate the need to enter (and the errors that can be introduced thereby) order and invoice information into the recipient's computer, and just as importantly can save considerable time in getting customer orders into the supplier's system.

5. Most common are purchase orders and invoices; also included in the current definition, though not necessarily complete in the standard, are P.O. change orders, Inventory Advice, Product Transfer and Resale Report, Price Sales Catalog, and Ship Notice/ Manifest.

6. The layers are: outer envelope (message control), functional group (group of like documents), transactions set (individual document within a group), and segments (parts of the document).

Answers to Chapter 11 Review Questions

1. Word processing is most often found on PC's, although it can reside on large systems.

2. Preserves images and makes accessible along with associated data such things as photographs, signed documents, and handwritten items.

3. The generic term database only means an organized repository of data elements. Database Management System (DBMS) packages, as are readily available in the marketplace today, offer convenient nonprogrammer access to data storage, retrieval, and manipulation capabilities.

4. A Query function is a utility that allows ad hoc retrieval of information without the need for writing programs.

5. E-Mail offers a vehicle for handling interpersonal and inter-departmental communications efficiently and effectively. Effective communications is an important contributor to efforts to develop cooperation and coordination.

6. Scheduling a meeting for a number of busy people can be done easily if their calendars are all accessible on a system. Ask the system when they are all available, and let *it* do the comparing!

7. A danger with PC-based spreadsheets in an environment with a centralized system or database is that the user can download information into his/her spreadsheet and manipulate it, but if he/she saves it for future use, his/her starting point for further manipulation is out of sync with the centralized data. Multiple, different copies of the same data are unacceptable in a controlled data system. With a diskless PC, since data cannot be stored locally, a fresh copy of the source data is retrieved each time and synchronization is assured.

8. Dial-up lines can support 1200 baud data communications. Newer modem technology has pushed this limit to 2400 baud.

9. Broadband cabling systems can support multiple simultaneous uses. This category includes coaxial cable and fiber-optic cables.

Acronyms

In our efforts to communicate efficiently with our peers, we tend to develop a shorthand in our own technical specialty that, unfortunately, tends to exclude the uninitiated. Most, but not all, of these acronyms appear in this book. They are commonly used in the manufacturing and CIM communities.

ABC Ranking of items in decreasing order of value or importance with assignment to categories (Pareto analysis)

Glossary

AI Artificial Intelligence

AIAG Automotive Industry Action Group

AIX IBM's version of the UNIX operating system

AGV Automated Guided Vehicle

ANSI American National Standards Institute

APICS American Production and Inventory Control Society

ARPA Advanced Research Projects Agency

ASQC American Society for Quality Control

AS/RS Automated Storage and Retrieval System

BOM Bill of Material

BRP Business Requirements Planning

CAD Computer Aided Design (Drafting)

CADD Computer Aided Design and Drawing (Drafting)

CAE Computer Assisted Engineering

CAM Computer Assisted Manufacturing

CAPP Computer Assisted Process Planning

CCITT International Telephone and Telegraph Consultative Committee

CIM Computer Integrated Manufacturing

CISC Complex Instruction Set Computer (opposite of RISC)

CMLT Cumulative Material Lead Time

CMfgLT Cumulative Manufacturing Lead Time

CMM Coordinate Measurement Machine

CNC Computer Numerical Control
CRP Capacity Requirements Planning
CRT Cathode Ray Tube (common name for a computer terminal)
CSMA/CD Carrier Sense Multiple Access with Collision Detection
DBMS Database Management System
DEC Digital Equipment Corporation
DNC Direct (or Distributed) Numerical Control
DP Data Processing
DRP Distribution Requirements (or Resource) Planning
EAN European Article Number (also called World Product Code or
      International Article Number)
EDI Electronic Data Interchange
EMI Engineering/Manufacturing Interface (IBM standard)
EOQ Economic Order Quantity
FLOPS Floating Point Operations Per Second
FMC Flexible Manufacturing (or Machining) Cell
FMS Flexible Manufacturing System
FTAM File Transfer Access Method
FTP File Transfer Protocol (part of TCP/IP)
GPIB General Purpose Interface Bus (same as IEEE-488)
GT Group Technology
HDLC High Level Data Link Control
HP Hewlett-Packard
IBM International Business Machines
ID Identification
IEEE Institute of Electrical and Electronics Engineers
IGES Initial Graphics Exchange Specification
ISDN Integrated Services Digital Network
ISO International Organization for Standardization
JIT Just-In-Time
k Thousand
kbps Thousand bits per second
LAN Local Area Network
MAD Mean Absolute Deviation
MAP Manufacturing Automation Protocol
MIPS Millions of Instructions Per Second
MIS Management Information System(s) or Services

MODEM Modulator/Demodulator

MPS Master Production Schedule

MRP Material Requirements Planning

MRP II Manufacturing Resource Planning

MS-DOS MicroSoft Disk Operating System (PC operating system)

NBS National Bureau of Standards

NC Numerical Control

OCR Optical Character Recognition

OP Order Point (inventory replenishment system)

OSF Open Software Foundation

OSI Open Systems Interconnect

PC Personal Computer, IBM or compatible (also Production Control or Programmable Controller)

PC-DOS Personal Computer Disk Operating System (same as MS-DOS, written by MicroSoft and marketed by IBM)

PDES Product Data Exchange Standard

PLC Programmable Logic Controller

PO Purchase Order

QA Quality Assurance

QC Quality Control

RCCP Rough Cut Capacity Planning

RDBMS Relational Database Management System

RISC Reduced Instruction Set Computer

RRP Resource Requirements Planning

RS-232, RS-422 Serial communications standards

SCADA Supervisory Control And Data Acquisition

SD Standard Deviation

SDLC Synchronous Data Link Control

SME Society of Manufacturing Engineers

SMED Single Minute Exchange of Dies

SMTP Simple Mail Transfer Protocol

SNA Systems Network Architecture (IBM standard)

SPC Statistical Process Control

STEP Standard for Exchange of Product (data)

SQC Statistical Quality Control

TCP/IP Transmission Control Protocol–Internet Protocol

TQC Total Quality Control

TQM Total Quality Management
UCC Uniform Commercial Code or Uniform Code Council
UNIX The AT&T operating system most often associated with work-station computers
UPC Universal Product Code
VAX Digital Equipment Corporation (DEC) system family
VMS A DEC operating system
WIP Work In Process
XENIX A version of UNIX
ZI Zero Inventory

Index